AF746102

Robotics Software Design and Engineering

Edited by Alejandro Rafael Garcia Ramirez and Augusto Loureiro da Costa

Published in London, United Kingdom

IntechOpen

Supporting open minds since 2005

Robotics Software Design and Engineering
http://dx.doi.org/10.5772/intechopen.91602
Edited by Alejandro Rafael Garcia Ramirez and Augusto Loureiro da Costa

Contributors
Trung Nguyen, Ngoc-Tam Bui, Edi Moreira M. de Araujo, Alejandro Rafael Garcia Ramirez, Augusto Loureiro Da Costa, Pingguo Huang, Yutaka Ishibashi, Ricardo Sanz, Esther Aguado, Diego Stéfano Fonseca Ferreira, Wagner Luiz Alves De Oliveira, Manuel Arturo Rendón Maldonado, Shiraishi Yoichi, HaoHao Zhang, Kazuhiro Motegi, Amar Nath, Rajdeep Niyogi, Luca Muratore, Nikolaos G. Tsagarakis, Arturo Laurenzi, Van-Tinh Nguyen

© The Editor(s) and the Author(s) 2021
The rights of the editor(s) and the author(s) have been asserted in accordance with the Copyright, Designs and Patents Act 1988. All rights to the book as a whole are reserved by INTECHOPEN LIMITED. The book as a whole (compilation) cannot be reproduced, distributed or used for commercial or non-commercial purposes without INTECHOPEN LIMITED's written permission. Enquiries concerning the use of the book should be directed to INTECHOPEN LIMITED rights and permissions department (permissions@intechopen.com).
Violations are liable to prosecution under the governing Copyright Law.

CC BY

Individual chapters of this publication are distributed under the terms of the Creative Commons Attribution 3.0 Unported License which permits commercial use, distribution and reproduction of the individual chapters, provided the original author(s) and source publication are appropriately acknowledged. If so indicated, certain images may not be included under the Creative Commons license. In such cases users will need to obtain permission from the license holder to reproduce the material. More details and guidelines concerning content reuse and adaptation can be found at http://www.intechopen.com/copyright-policy.html.

Notice
Statements and opinions expressed in the chapters are these of the individual contributors and not necessarily those of the editors or publisher. No responsibility is accepted for the accuracy of information contained in the published chapters. The publisher assumes no responsibility for any damage or injury to persons or property arising out of the use of any materials, instructions, methods or ideas contained in the book.

First published in London, United Kingdom, 2021 by IntechOpen
IntechOpen is the global imprint of INTECHOPEN LIMITED, registered in England and Wales, registration number: 11086078, 5 Princes Gate Court, London, SW7 2QJ, United Kingdom
Printed in Croatia

British Library Cataloguing-in-Publication Data
A catalogue record for this book is available from the British Library

Additional hard and PDF copies can be obtained from orders@intechopen.com

Robotics Software Design and Engineering
Edited by Alejandro Rafael Garcia Ramirez and Augusto Loureiro da Costa
p. cm.
Print ISBN 978-1-83969-291-8
Online ISBN 978-1-83969-292-5
eBook (PDF) ISBN 978-1-83969-293-2

For EU product safety concerns:
IN TECH d.o.o., Prolaz Marije Krucifikse Kozulić 3, 51000 Rijeka, Croatia, info@intechopen.com or visit our website at intechopen.com.

We are IntechOpen, the world's leading publisher of Open Access books
Built by scientists, for scientists

5,400+
Open access books available

133,000+
International authors and editors

165M+
Downloads

156
Countries delivered to

Our authors are among the
Top 1%
most cited scientists

12.2%
Contributors from top 500 universities

WEB OF SCIENCE™

Selection of our books indexed in the Book Citation Index
in Web of Science™ Core Collection (BKCI)

Interested in publishing with us?
Contact book.department@intechopen.com

Numbers displayed above are based on latest data collected.
For more information visit www.intechopen.com

Meet the editors

Alejandro Rafael Garcia Ramirez is Professor of Applied Computing at Universidade do Vale de Itajai, Brazil, and a member of the Mechanical and Computer Departments. He received his certificate of higher education in Electronic Engineering from Instituto Superior Politecnico José Antonio Echeverria, Havana, Cuba, in 1989. He received his Ph.D. in Electrical Engineering from Universidade Federal de Santa Catarina, Florianopolis, Brazil, in 2003. He has published more than 135 papers in local and international peer-reviewed journals and scientific conferences in the fields of robotics, technologies for education, embedded systems, and assistive technologies. Since 2010, he has visited Cuba as a researcher conducting and supervising an international cooperation project between Universidade of Havana and Universidade do Vale de Itajai. He is a member of the ad-hoc advisory committee of Coordination for the Improvement of Higher Level or Education Personnel (CAPES) and National Council for Scientific and Technological Development (CNPq) and serves on the scientific committee of several journals and scientific conferences. He completed a CNPq fellowship in 2013. He has also received several awards, including the Distinguished Scholars Award from the Instituto Superior Politecnico José Antonio Echeverria, Havana, Cuba, in 1989 and the Entrepreneurship Incentive Program Award from Fundação de Amparo à Pesquisa e Inovação do Estado de Santa Catarina (FAPESC), Brazil, in 2012.

Augusto Loureiro da Costa is a full professor at the Department of Electrical and Computer Engineering, Polytechnic School of the Federal University of Bahia (UFBA), Brazil. He obtained a Ph.D. in Electrical Engineering from the Federal University of Santa Catarina, Brazil, in 2001, held in sandwich mode with the University of Karlsruhe, Germany. He obtained a master's in Electrical Engineering from the Federal University of Santa Catarina in 1997. He also obtained a degree in Civil Engineering from UFBA in 1993. Dr. Loureiro da Costa was a visiting professor at the University of Pennsylvania, General Robotic Automation and System of Perception Laboratory (GRASP) Lab during 2012–2013. His research interests include multi-robot systems, cooperative robotics, autonomous robots, autonomous agents, and multi-agent systems. He is a coordinator of the Intelligent Systems Area of the Sociedade Brasileira de Automática. From 2008 to 2010, he was also coordinator of the Special Committee on Artificial Intelligence of the Brazilian Computer Society as well as coordinator of the Collegiate Course on Computer Engineering at UFBA. He was head of the Computer and Department of Electrical Engineering at UFBA in 2010–2012.

Contents

Preface

Robotics Software Design and Engineering presents results from the scientific community in robotics, artificial intelligence, embedded systems, multi-robot systems, and Industry 4.0, among other interesting topics. The chapters explore robotics applications and related challenges, including interaction protocols, robot software frameworks, robots with residual degrees of freedom, evolution of algorithms for Quality of Service (QoS) control, autonomous robots, system-on-a-chip (SoC) designs, interfaces for simulation and control development, autonomous driving, and autonomous cooperative transportation.

Alejandro Rafael Garcia Ramirez
Department of Applied Computing,
University of Vale de Itajai,
Itajaí, Brazil

Augusto Loureiro da Costa
Department of Computer and Electrical Engineering,
Polytechnic School of the Federal University of Bahia (UFBA),
Salvador, Brazil

Chapter 1

XBot: A Cross-Robot Software Framework for Real-Time Control

Luca Muratore, Arturo Laurenzi and Nikos G. Tsagarakis

Abstract

The widespread use of robotics in new application domains outside the industrial workplace settings requires robotic systems which demonstrate functionalities far beyond that of classical industrial robotic machines. The implementation of these capabilities inevitably increases the complexity of the robotic hardware, control a and software components. This chapter introduces the XBot software architecture for robotics, which is capable of Real-Time (RT) performance with minimum jitter at relatively high control frequency while demonstrating enhanced flexibility and abstraction features making it suitable for the control of robotic systems of diverse hardware embodiment and complexity. A key feature of the XBot is its cross-robot compatibility, which makes possible the use of the framework on different robots, without code modifications, based only on a set of configuration files. The design of the framework ensures easy interoperability and built-in integration with other existing software tools for robotics, such as ROS, YARP or OROCOS, thanks to a robot agnostic API called XBotInterface. The framework has been successfully used and validated as a software infrastructure for collaborative robotic arms as KUKA lbr iiwa/lwr 4+ and Franka Emika Panda, other than humanoid robots such as WALK-MAN and COMAN+, and quadruped centaur-like robots as CENTAURO.

Keywords: software architecture for robotics, real-time control, cross-robot framework, humanoid robotics, hardware abstraction layer, XBot, ROS

1. Introduction

Nowadays effective robotic solutions targeting new applications outside the traditional industrial environment, are supposed to operate in partially known spaces with unforeseen uncertainty and increased variability in the application tasks. Hence, to be effective, they have to adapt rapidly and seemly their functionalities in these demands, leading to an increase of the complexity in each layer of the robotic system, from the hardware to the high level control.

To tackle this, several software frameworks for robotics have been developed in the past twenty years, as stated in [1], aiming to provide flexible infrastructures, which not only permit the seamless integration of new functionalities and interfaces in the robotic system, but also ensure standardization, easy tracking and maintenance of the software development, despite the increased complexity. Apart from dealing with the software complexity, these frameworks have to provide hard Real-Time (RT) performance, ensuring predictable response times [2] as required in critical tasks when robots need to perform in autonomous mode, responding to

 IntechOpen

disturbances and interacting with the physical environment during the execution of a task. Thus, a vital feature of a software framework for robotics is the Real-Time safeness and scheduling, essential for precise robot control, especially when dealing with high frequency and low jittering control cycles.

Furthermore, a software middleware needs to abstract the complex hardware (e.g. actuators and sensors) of the robot providing an easy-to-use, standardized Application Programming Interface (API). As a matter of fact, a robot can be considered a distributed system composed of a set of hardware devices communicating through a fieldbus. The fast prototype and development of control and application software which can be shared, ported and reused in various robotic platforms with minimum effort, is another fundamental requirement for the software architecture. An important component needed to achieve this goal is the Hardware Abstraction Layer (HAL), which can be incorporated to mask the physical hardware differences and limitations (e.g. control frequency, kinematics/dynamics model, actuators type and size, sensors, etc) varying from one robot to another. The HAL can provide a relatively uniform abstraction layer that assures portability and code reuse: it permits the development of control modules that can be easily ported from one robot to another.

The existing robotics software frameworks address different needs and requirements, therefore one of the key aspects for a brand-new middleware is the interoperability with well-known and established robotic software platforms. Interoperability should ideally allow users to execute existing software without the necessity of (i) changing the current code and (ii) writing hand-coded "bridges" for each use case [3].

In this chapter the *XBot* software framework is presented. The development of the *XBot* was driven by the need to provide a software framework that abstracts the diverse variability of the robotic hardware (effectively becoming a cross robot platform framework), providing deterministic hard Real-Time (RT) performance, incorporating interfaces that permits it to integrate state of art robot control frameworks and achieve enhanced flexibility through a plug-in architecture.

2. Related works

In this section the state of the art of robotic software architectures will be analyzed.

In [4] the low level control framework, called **OROCOS** (Open Robot Control Software), is introduced, which provides a set of components for RT control of robotic systems. **OROCOS** relies on the Common Object Request Broker (CORBA) architecture, that allows inter-process and cross-platform interoperability for distributed robot control. Depending on any Inter-Process-Communication (IPC) framework can critically increase the complexity of the software platform. Despite *OROCOS* is used in a fair number of robotics projects, the framework maintenance as well as the community looks not being very active anymore[1].

Very similar to OROCOS is **OpenRT-M** [5], developed in Japan from 2002 under NEDO's (New Energy and Industrial Technology Development Organization) "Robot challenge program". It is based on CORBA, so similar considerations as for OROCOS can be made with respect to the software complexity; moreover part of OpenRT-M documentation is in Japanese.

---

[1] In particular we refer to the discontinuity in maintaining the framework under last versions ($\geq$3.X) of Ubuntu Xenomai, where the OROCOS porting is still experimental.

Framework	RT	HAL	IPC Complexity	Ready-to-Use	Community
ROS	No	Yes	Low	Yes	Big and active
YARP	No	Yes	High	Yes	Medium and active
OROCOS	Yes	No	Very High	Yes	Medium and inactive
OpenRT-M	Yes	Yes	Very High	Yes	Part of docs in Japanese
ROS 2	Yes	Yes	High	No	Small and Active
PODO	Yes	Yes	Very High	No, not available	KAIST group only
IHMC OpenJDK	Yes, low performance	No	Medium	Yes	Small
XBot	*Yes*	*Yes*	*Low*	*Yes*	*Small*

Table 1.
Summary of the features of the available software frameworks for robotics: The XBot was developed from scratch given the limitations and the missing features of the presented existing framework.

YARP (Yet Another Robot Platform) [6] and **ROS** (Robot Operating System) [7] are popular component-based frameworks for IPC that do not guarantee RT execution among modules/nodes. It is essential for a robotic system to have a component responsible for the RT control of the robot, making these frameworks only viable as external (high-level) software frameworks. It is worth to mention that a new *ROS* version, called **ROS 2**[2] has been released: it is still in an early stage development phase, so it cannot be used in real-world scenario[3].

PODO [8], is the framework used by KAIST in HUBO during the DRC (Darpa Robotics Challenge) Finals. Its control system has RT control capabilities and its inter-process communication facilities are based on POSIX IPC; moreover it uses a shared memory system called MPC to exchange data between processes in the same machine. This heterogeneous system has the potential to cause confusion as it is unclear which architectural style must be used to communicate with a specific component [9].

In [10] an RT architecture based on OpenJDK is introduced (used by IHMC during the DRC Finals). Nevertheless, to their own admission [11], none of the commercially available implementations of the Java Real Time Specification had the performance required to run their controller. Existing Real-time Java Support is insufficient.

Considering the above limitations, summarized in **Table 1**, the *XBot* [12–14] was developed from scratch, in order to have a reliable RT control framework with HAL support and without depending on complex IPC frameworks.

3. XBot framework

The development of *XBot* was driven by the need to provide a software infrastructure that abstracts the diverse variability of the robotic hardware (effectively becoming a cross-robot framework), provides deterministic hard Real-Time (RT) performance, incorporates interfaces that permit its integration with state of art robot control frameworks and achieves enhanced flexibility through a plug-in architecture.

[2] https://index.ros.org/doc/ros2/
[3] https://index.ros.org/doc/ros2/Features/

In the next sections, the *XBot* design goals and the software architecture insights are going to be described following a bottom-up approach going from the hardware towards the high-level control of the robotic system.

3.1 Design goals

The considerations and limitations of the existing frameworks, described in detail in the previous section, motivated the development of the *XBot* framework bearing in mind that the design of a software platform, which lies at the foundations of such complex and diverse robotic systems, is the most crucial phase in the software development process. *XBot* was designed to be both an RT control system and a user friendly, flexible and reusable middleware for RT and non-RT control software modules. *XBot* was developed starting from the following design goals and features:

- **Hard RT control performance**: it must perform computation inside specific timing constraints with minimum timing jitter. There are several operating systems or platforms which support RT operation, including Windows CE, INtime, RTLinux, RTAI, Xenomai, QNX and VXWorks. Xenomai[4] [15], a Linux based Real-Time Operating System (RTOS) was selected to avoid a licensed product that does not give the possibility to modify and adapt the source code to fit it to the specifications of the system. Moreover Xenomai satisfies the requirements for extensibility, portability and maintainability as well as ensuring low latency as stated. in [16, 17].

- **High control frequency**: robotics applications may often require high frequency control loops, e.g. RT pattern generator for bipedal walking, impedance regulation controllers or force feedback modules.

- **Cross-Robot compatibility**: it should be possible to use it with any robot, without code modification. It is crucial to be able to reuse the software platform with different robots, or subsystems of the same robotic platform.

- **External Framework integration**: it should be possible to use *XBot* as a middleware for any kind of external software framework (RT or non RT) without tailored software or specific bridges for every different case.

- **Plug-in Architecture**: users and third parties should be able to develop and integrate their own modules. In a robotic system platform a highly expandable software structure is needed.

- **Light-weight**: small number of dependencies on other libraries, it should be easy to install and set up. It is expected to run *XBot* on embedded PCs with low performance requirements in terms of memory and CPU. Therefore, it should demonstrate a small footprint to avoid high CPU usage.

- **Simplicity**: it must be simple. Complex systems may have unneeded and over-engineered features. For robotics application full control over the software platform is required. *KISS* ("Keep It Simple, Stupid") principle is essential and unnecessary complexity should be avoided.

---

[4] https://xenomai.org/

- **Flexibility**: *XBot* has to be easily modified or extended to be used in systems and applications other than those for which it was specifically designed.

Finally, the *XBot* software framework was not developed to address the requirements of a specific robotic platform, instead its implementation is flexible, generic and cross-robot. Furthermore it does not directly depend on any existing software or control platform, but it provides to the user the functionality to easily integrate any RT or non-RT framework. To obtain the above features, a user of the *XBot* with a generic robotic platform to control, has to provide a set of configuration files, mainly related to the kinematics and dynamics of the robot, the control plugins to execute, the HAL implementation, the high level communication framework and the kinematic/dynamic engine to use.

3.2 Software architecture

As presented in **Figure 1**, the *XBot* software architecture is composed of different components, described in detail within the following sections. In particular the design choices, from a software engineer point of view, are the results of the design goals described in Section 3.1. To avoid scheduling issues and keep the complexity of the software infrastructure as low as possible only two RT threads and one non-RT thread are currently employed in the framework as presented in **Figure 2**. The RT layer contains the R-HAL (Robotics Hardware Abstraction Layer) to assure cross-robot compatibility and seamless porting of the higher level code from the simulation to the real robot. The communication mechanism employed in this layer is a shared memory one using basic synchronization methods (i.e. mutex and condition variables). On top of the R-HAL, the *Plugin Handler* is designed using a

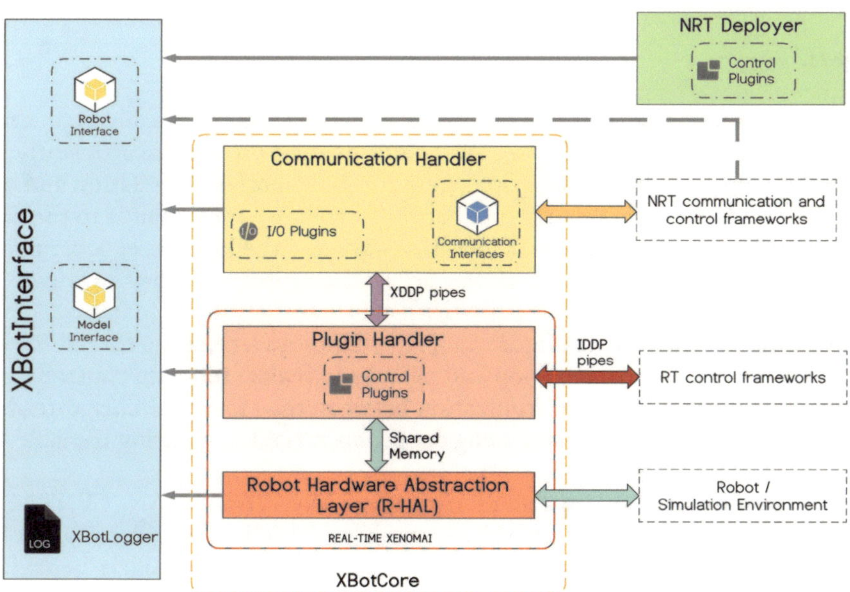

Figure 1.
XBot *software architecture: components overview and interaction. From the bottom the* R-HAL *and the plugin handler inside the RT Xenomai layer are presented, with the shared memory communication between them. The non-RT layer and the external software integration component of the* XBot *is represented by the* Communication Handler *which is able to communicate with the robot/simulation through a XDDP mechanism which assures lock-free IPC. Thanks to the* XBotInterface *on the left, all the layers of the framework have an uniform way (through an API) to send commands and receive state from the robot.*

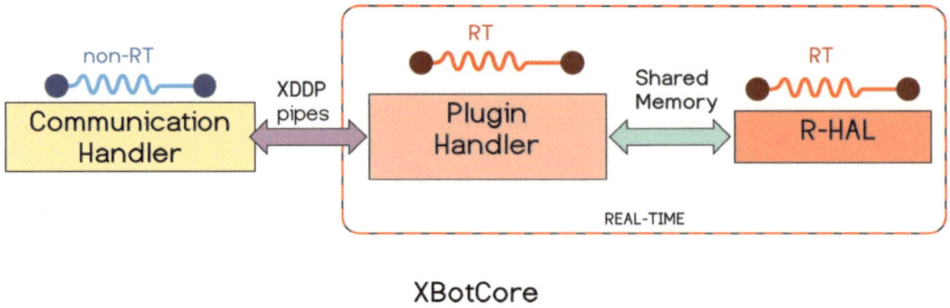

Figure 2.
XBot *threads structure and communication mechanisms: To keep the complexity of the framework low and assure full control over the software infrastructure, only two RT threads and one non-RT thread are currently employed.*

component-based software design paradigm by featuring a clear component concept (the plugin as a shared library) with well-defined structure and communication interfaces. In order to be able to assure external software integration, the communication with the non-RT layer, represented by a standalone component called `Communication Handler`, is implemented using a lock-free Inter Process Communication (IPC) mechanism based on XDDP (described in details in the following sections). The same concept is applied to communicate with other RT frameworks (e.g. OROCOS), using a mechanism based on IDDP. The idea of the *NRT Deployer* came from the need to have a behavior similar to the one of the *Plugin Handler* but completely in the non-RT layer. Finally a standard way of communicating with the robot regardless of its specific structure (humanoid, quadruped, manipulator, etc), and also independently of the particular software layer that the user wants to operate within, is provided by mean of the *XBotInterface*.

3.2.1 R-HAL

The Cross-Robot compatibility feature is achieved through the development of a suitable hardware abstraction layer [18], which enables the user to efficiently port and run the same control modules on different robots, both in simulation and on the real hardware platforms. The main goal of this software component is to provide an independent layer in between the robot hardware and the high-level control, enabling the seamless integration of new actuators, sensors or other hardware components.

Concerning the threads configuration, *XBot* employs a separate thread to execute the low-level robot control loop and permits to realize separate controllers with different frequencies. The synchronization between the *Plugin Handler* thread and the *R-HAL* thread is implemented using condition variables, assuring the safe access of the shared data structures.

XBot currently supports EtherCAT (for robots like WALK-MAN, CENTAURO and COMAN+), Ethernet (for COMAN), and KUKA LWR 4/KUKA LBR arm based robots [19–21]. The possibility to simulate the robot and its controllers behaviors prior to testing on the real hardware is essential, especially when dealing with complex robotic systems. To achieve this we provide an *R-HAL* implementation for the well known *Gazebo*[5] simulator environment **Figure 3**. In particular we rely on the *Gazebo* ModelPlugin class to be part of the Gazebo internal loop.

[5] http://gazebosim.org/

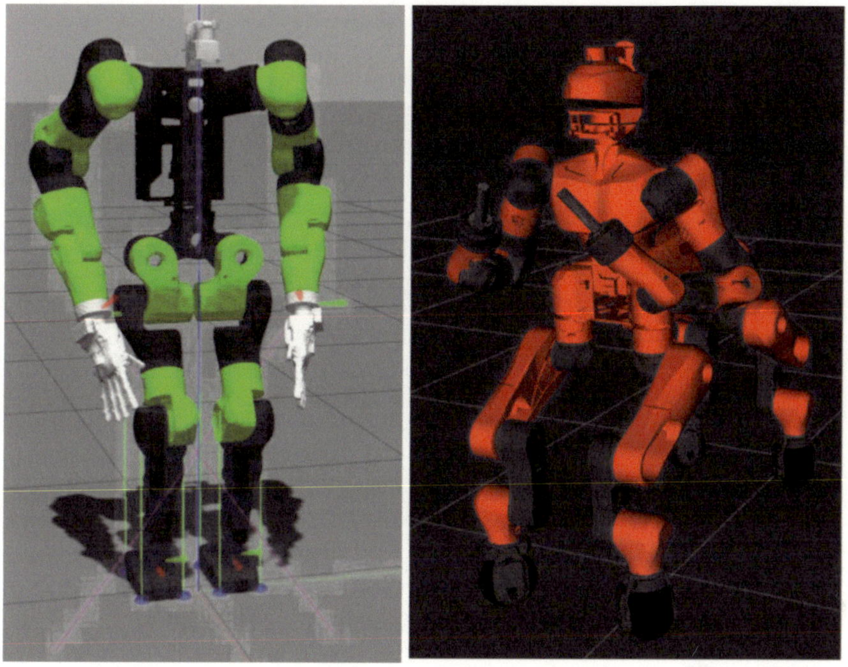

Figure 3.
COMAN+ robot controlled inside the gazebo simulator (left) and CENTAURO robot in RViZ (right): Both using two different implementations of the R-HAL *provided in the* XBot *software architecture.*

3.2.2 Plugin handler

The main component of the *XBot* architecture is called *PluginHandler* and it is represented in **Figure 1** with the dark pink color. The software design of this component relies on two core requirements for a robotic system (described in 3.1): the RT control and the highly expandable software structure. To achieve this the *PluginHandler* is implemented using a single RT thread running at high frequency (e.g. 1 kHz) and it is responsible for the following actions with the order they appear below: load the set of plugins requested by the user from a configuration file, initialize all the loaded plugins, and start them upon user request, execute the plugins that have been started sequentially, reload and reinitialize a plugin upon user request, close and unload all the loaded plugins. In **Figure 4**, the UML state diagram representing the life-cycle of a plugin is presented.

The Plugin implementation is compiled as a shared object library (.so). In details a Plugin is a simple class inherited from the abstract class `XBotControlPlugin`; this means that writing a Plugin is straightforward for the user, as the only need is to implement three basic functions:

- an *init_control_plugin()* function, which is called by the *PluginHandler* after the plugin is loaded/reloaded and is useful to initialize the variables of the Plugin

- a *control_loop()* function, which is called in the run loop of the *PluginHandler* after the plugin is started

- a *close()* function, which is called in the *PluginHandler* closing phase

After the design and the implementation of the latency-free, hard real-time layer the next significant feature is accompanied by the implementation of flexible

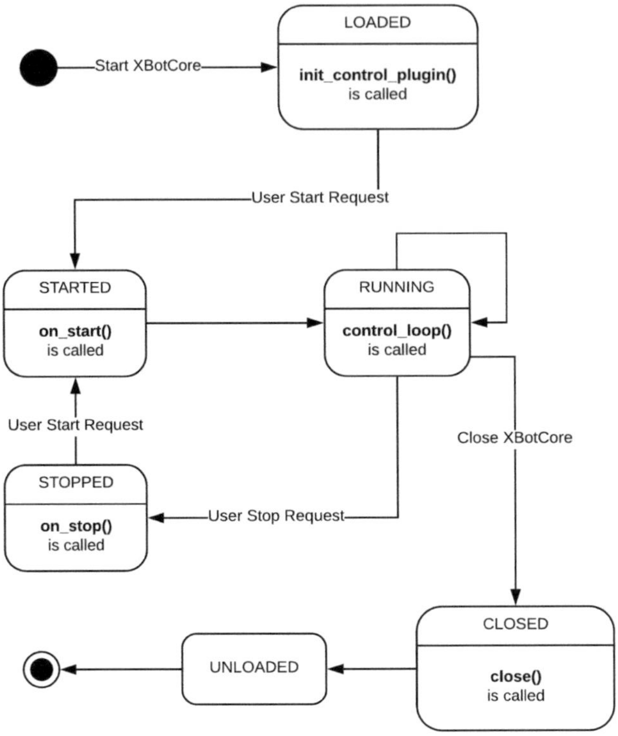

Figure 4.
UML state diagram showing a XBot *plugin life-cycle.*

interfaces, called *XBotInterface*, which permit our framework to integrate with state-of-art, widely spread robot control frameworks.

3.2.3 Communication handler

The above mentioned software components do not give the possibility to communicate with external modules/hosts outside the robot: for this purpose the software framework of a robotic system should incorporate a set of non-RT threads that permit the communication of the system with remote pilot stations or cloud services. *XBot* provides this with the implementation of the `Communication Handler` component (represented in **Figure 1** with the yellow color) that is a non-RT thread exploiting an XDDP (Cross Domain Datagram Protocol) handler with the ready-to use *XBot* non-RT API for a set of components called `CommunicationInterface(s)`. The non-RT API uses XDDP Xenomai pipes to achieve asynchronous communication between RT and non-RT threads. A lock-free inter-process communication (IPC) is employed to permit the RT control threads to exchange messages with the non-RT communication threads without any context switch. The execution loop of the `Communication Handler` thread is responsible for updating the internal robot state using the XDDP pipe with the non-RT robot API, sending the robot state to all the communication frameworks implemented as `CommunicationInterface(s)`, receiving the new reference from the "master" `CommunicationInterface` (to avoid having multiple external frameworks commanding the robot) and finally for sending the received reference to the robot using the XDDP non-RT robot API.

It is relatively straightforward to add a new `CommunicationInterface` in the framework: *XBot* provides built-in support for YARP and ROS communication frameworks, meaning that the end-users has YARP control board wrappers / analog sensors and ROS joint state / command messages already available. Interoperability for YARP/ROS framework and *XBot* is one of the key feature offered by the `Communication Handler`.

In the specific ROS case, *XBot* provides two families of interfaces:

1. a joint space interface, consisting of standard ROS topics that are advertised/ subscribed by the *Communication Handler* itself in order to publish the robot state (including sensors) and accept commands

2. a set of tools for using a subset of ROS inter-process communication capabilities from the RT domain

ROS-powered robots expose to their users an interface that is mainly based on topics. For instance, the robot joint state is usually published to a `/joint_state` topic through `sensor_msgs/JointState` messages. The same kind of message can be published by the user on a `/command` topic in order to control the robot.

Inside *XBot* a similar interface to the ROS middleware is offered, the only difference lies in the message type being used. Broadly speaking, *XBot* uses an extended joint state message that make it possible to perform more flexible control of the robot than is allowed with standard ROS.

The *XBot* framework provides also the integration with any external RT software framework (e.g. OROCOS) thanks to the use of the IDDP (Intra Domain Datagram Protocol) pipes for the RT inter-process communication.

Inside the `Communication Handler` component, the *XBot* framework provides the user with the possibility of running non-RT plugins that are useful for performing Input/Output operation from the non-RT layer to the RT layer. The so called `IOPlugins` are very similar to the Plugin used in the RT layer, in fact the implementation is compiled as a shared object library (.so). In detail, an `IOPlugin` is a simple class inheriting from the abstract class `XBot::IOPlugin`, which gives to the user simple access to the shared memory component and the pipes to communicate between the non-RT layer and the RT one. As for the standard Plugin, *XBot* provides a ready-to-use skeleton (simple script to run) for the user.

4. Experimental validation

In this section the results of the validation of the overall framework is going to be presented with particular focus on the flexibility in terms of integration with different robots and external software frameworks, and also on the overhead introduced by the *XBot* while assuring predictable response time at high frequency (i.e. 1 kHz) control loop.

4.1 Experimental setup

To evaluate the performance of the *XBot* software platform, two sets of experiments were performed: in the experiment **set 1** the WALK-MAN [22, 23] robot was used, a humanoid with 33 Degree-Of-Freedom, 4 custom F/T sensors and 1 VN-100 imu. The WALK-MAN vision system is composed of a CMU Multisense-SL sensor that includes a stereo camera, a 2D rotating laser scanner, and an IMU. The robot control modules were based on GYM [24] (Generic Yarp Module), a component model to easily develop software modules for robotics leveraging the YARP ecosystem: YARP Based Plugins for Gazebo Simulator were used to validate the control

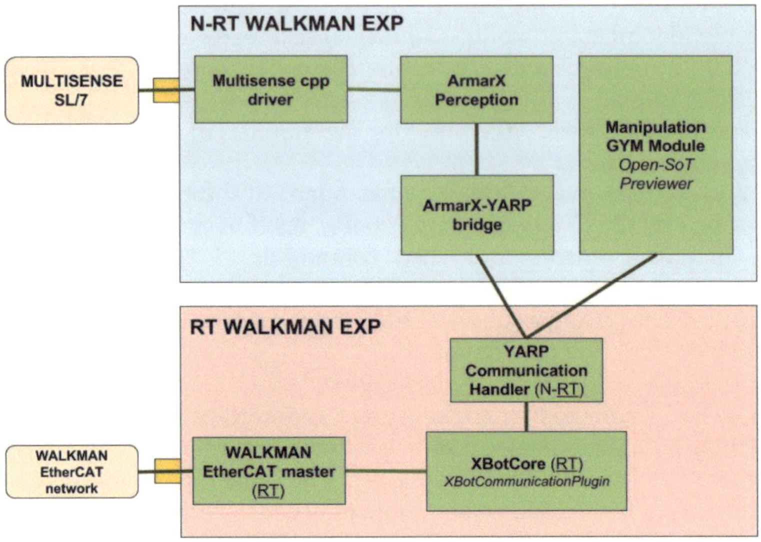

Figure 5.
XBot *validation experiment set 1 software components: How they are allocated in the two WALK-MAN embedded PCs (i.e. N-RT WALKMAN EXP and RT WALKMAN EXP).*

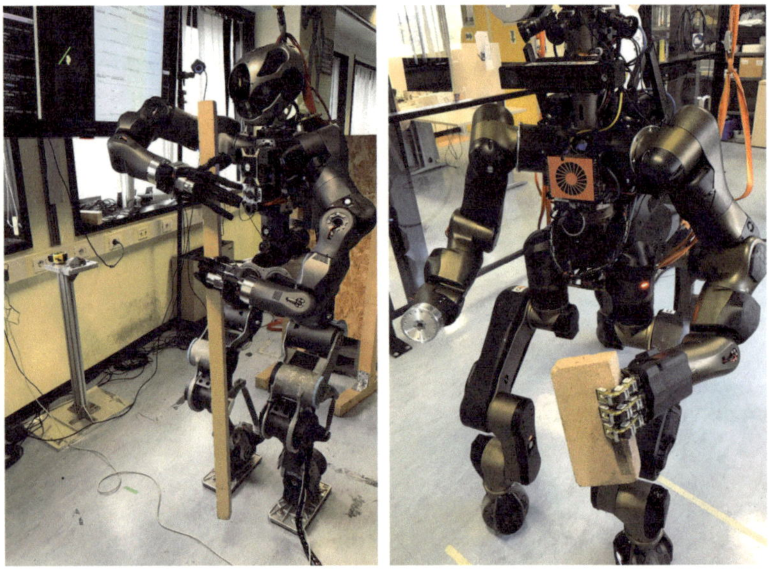

Figure 6.
XBot *validation experiment. On the left* ***set 1*** *setup: WALK-MAN needs to remove a set of objects in order to perform the task of turning the valve. On the right CENTAURO bi-manual robot platform used in experiment* ***set*** *2.*

modules in simulation. Whole-body control and inverse kinematics are solved through the OpenSoT control framework [25]. **Figure 5** reports a representation of the software components in use for experiment **set 1**.

In the **set 1** evaluation different high-level software frameworks were successfully integrated on top of *XBot*: *ArmarX* [26] perceptual pipeline for hierarchical affordance extraction [27], OpenSoT previewer based on the *MoveIt! ROS* library

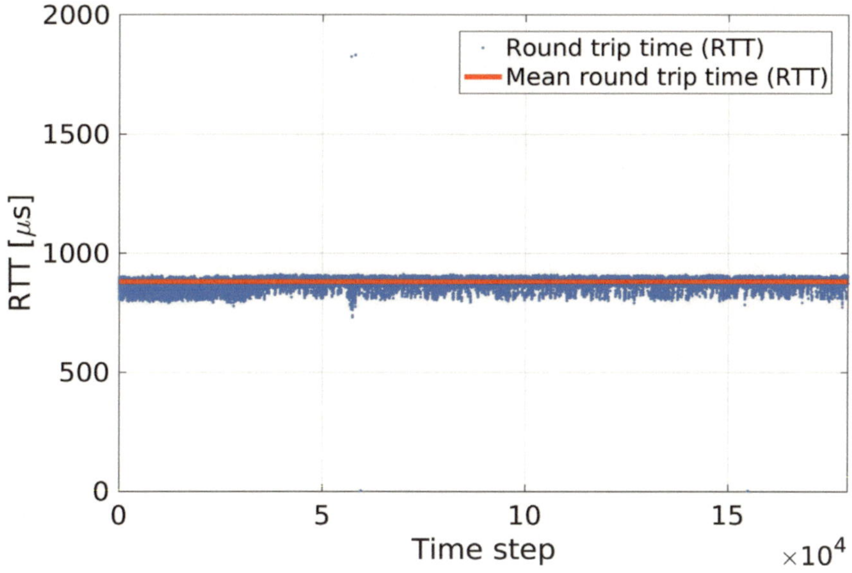

Figure 7.
Experiment ***set 1****: WALK-MAN EtherCAT slaves RTT measured by EtherCAT master during manipulation actions:* XBot *assures always a control period below 1000 μs while providing integration with external software frameworks as described in* ***Figure 5****.*

for motion feasibility analysis and collision checking and a manipulation GYM module, OpenSoT based, using the *YARP* communication framework.

The **set 1** experiments were carried out in a DRC-inspired scenario targeting the removal of debris in front of a valve. In **Figure 6** the experimental setup is shown.

In [28], ArmarX was integrated with the robot software environment YARP taking advantage of the built-in YARP `CommunicationInterface` for the external software framework integration with *XBot*.

In the experiment **set 2** an RT end-effector Cartesian Control on two different robots was performed: the aforementioned WALK-MAN and CENTAURO (in **Figure 6**). CENTAURO [20, 29, 30] upper body is a high performance human size and weight compatible bi-manual manipulation platform with 15 DOFs. Each arm has 7 DOF and the trunk has 1 DOF that permits the yaw motion of the entire upper body and extends the manipulation workspace of the robot.

In the **set 2** experiments two RT Plugins were used: the first one, called `IKCommunication` to receive the end-effector pose from the `Communication Handler` (with the built-in ROS `CommunicationInterface`) through the XDDP pipes and OpenSoTRTIK to solve the inverse kinematics. The evaluation was focused on the overhead introduced by the `IKCommunication` RT plugin that exploits two communication mechanism offered by *XBot*: XDDP to receive the data from the non-RT layer and `XBotSharedMemory` to communicate these data to the other RT plugin (`OpenSoTRTIK`).

4.2 Results

In the **set 1**, *XBot* performance in terms of control period of the RT plugin `XBotCommunicationPlugin` and CPU usage were analyzed: during the experiments, each millisecond, all the data flowing from/to the R-HAL were recorded, using the *XBot* RT low-level logging tools.

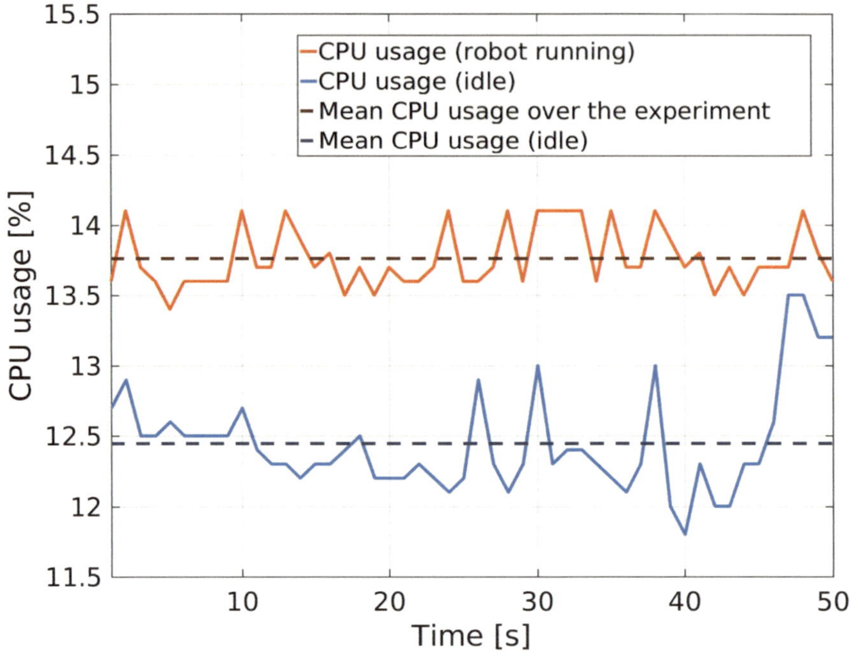

Figure 8.
*Experiment **set 1**: XBot CPU core usage comparison: Robot idle vs. robot running the experiments. The difference between the two bold lines represents the actual overhead introduced by the middleware when executing the control modules described in the experimental setup for **set 1**.*

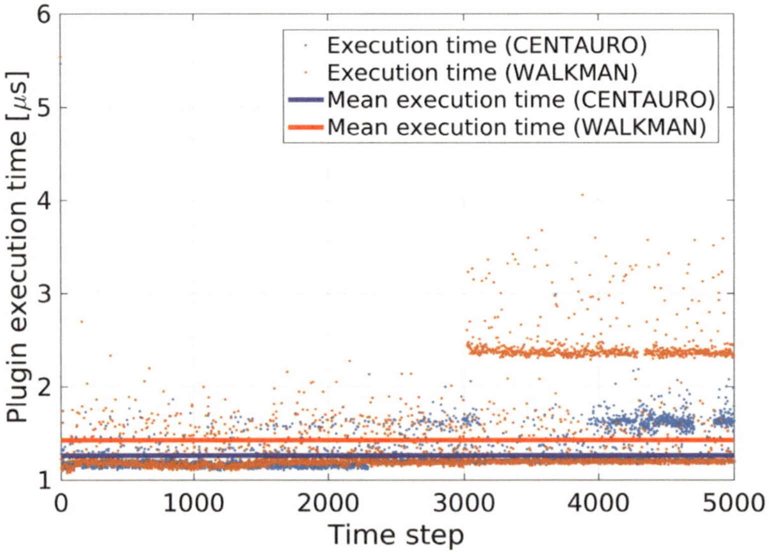

Figure 9.
*Experiment **set 2**: Communication overhead (RT - RT and N-RT - RT) introduced by XBot. Experiment results on both WALK-MAN and CENTAURO are shown.*

In **Figure 7** the RTT (Round Trip Time) measured by the EtherCAT master implementation of the R-HAL during the **set 1** experiments in the worst-case scenario is shown, i.e. while the robot was performing the manipulation actions: it is

Figure 10.
XBot *framework usage examples: WALK-MAN robot in Pisa (top left), CENTAURO robot untethered and outdoor (bottom left), and the COMAN and its scaled-up version COMAN+ humanoid robots shaking hands (right).*

clear that the mean control period is below the 1000 μs (i.e. 1 kHz control frequency) even if the RT system is communicating with the high-level software components through *XBot* the built-in YARP `CommunicationInterface` non-RT threads. Only two of the RTT measurement (over 200000) were above the requested control period because of missing PDO (Process Data Objects) round in the beckhoff[6] chip responsible for the EtherCAT communication.

In **Figure 8** a comparison is presented between *XBot* CPU usage while the robot is idle (i.e. not moving, nor communicating with external software frameworks) and when the **set 1** manipulation experiments are running: the CPU core usage overhead introduced by *XBot* when the robot is performing the manipulation task as described above, is only 1.2% (on average). Furthermore it is clear that the CPU usage of *XBot* is very low (always ranging from 11.7% to 14.2%).

In the **set 2** the focus was placed on the communication overhead introduced by *XBot*: both the XDDP pipes (communication between RT and non-RT layers) and the `XBotSharedMemory` (RT plugins communication) are taken into account. As shown in **Figure 9** the mean execution time of the `IKCommunication` RT plugin is around 1.2 μs for both WALK-MAN and CENTAURO experiments. This means that it is possible to send end-effector reference poses and receive back the robot state from a non-RT framework, while controlling the robot (at 1 kHz in the experiments) using a RT plugin implementing the IK (`OpenSoTRTIK` in the experiments), with negligible overhead (**Figure 10**).

---

6 https://www.beckhoff.it/

5. Conclusions

In this chapter the *XBot*[7] RT software architecture was presented. It provides to the users a software infrastructure which can be used with any robotic systems enabling fast and seamless porting of the code from one robot to the other, requiring no code changes, assuring flexibility and reusability. The implementation of the framework ensures easy interoperability and built-in integration with other existing software tools for robotics, such as ROS, YARP or OROCOS. The component-based development of the *XBot* includes a Robotic Hardware Abstraction Layer (R-HAL) interface and a set of ready-to-use tools to control robots either within a simulation environment or the real hardware.

The framework has been successfully used an validated as a main software infrastructure (**Figure 5**) for humanoid robots such as WALK-MAN (result of WALK-MAN EU FP7 project[8], notably *XBot* received the EU innovation radar award in this context[9].) and COMAN+ (result of COGIMON EU H2020 project[10]) or for quadruped centaur-like robots as CENTAURO (result of the CENTAURO EU H2020 project[11]). Moreover the cross-robot functionality has been exploited to develop both RT and non-RT control modules not only for the above mentioned robots, but also for commercial robotic systems such as KUKA LBR, KUKA 4+ or Franka Emika Panda, or other humanoid robots like COMAN or iCub.

Regarding the simulation part, *XBot* enables the direct porting of the control modules tested in the simulator to the real hardware using the same interfaces and without requiring any code modifications. The built-in simulator supported in the framework is Gazebo, but there is the option to support other simulation environments (as it happened inside the CENTAURO H2020 project with the VEROSIM simulator[12]).

XBot currently relies on a dual-kernel approach using Xenomai, which performs better than PREEMPT_RT[13], both in terms of system predictability and absolute latencies. Nevertheless Xenomai in the long term can introduce disadvantages by making the software development more complex, which means harder maintainability and lower portability.

Further development of the framework will target to provide synchronized distributed execution of multiple RT threads in multiple computational units. In fact currently the *Plugin Handler* is only able to execute a set of plugins in sequence, without any concurrency. This makes the maintenance of the framework easier, but restricts the performance in terms of computation power. Moreover the current architecture is characterized by a unique point of failure since both the *R-HAL* thread and the *Plugin Handler* (which executes RT plugins) thread run in the same process. In fact, there is the possibility that a misbehaving RT plugin might cause memory corruption, or crash altogether, causing also the *R-HAL* to crash. Currently only expert users are allowed to load their RT plugins in the *Plugin Handler*, but it is desirable to eventually separate the *R-HAL* and *Plugin Handler* either in two different processes or in two different machines to improve isolation.

---

7 https://github.com/ADVRHumanoids/XBotControl

8 https://www.walk-man.eu/

9 https://www.innoradar.eu/innovation/30632

10 https://cogimon.eu/

11 https://www.centauro-project.eu/

12 https://www.verosim-solutions.com/en/

13 PREEMPT_RT was introduced to have RT capabilities in the Linux kernel avoiding the adoption of a dual-kernel.

Acknowledgements

This project has received funding from the European Union's Horizon 2020 research and innovation programme under grant agreement No. 101016007 CONCERT, and the European Union's Horizon 2020 Research and Innovation Programme (H2020-ICT-2019-2/ 2019-2023) under grant agreement No. 871237 SOPHIA and the Italian Fondo per la Crescita Sostenibile – Sportello "Fabbrica intelligente", PON I&C 2014 - 2020, project number F/190042/01-03/X44 RELAX.

Author details

Luca Muratore*, Arturo Laurenzi and Nikos G. Tsagarakis
Humanoids and Human Centered Mechatronics (HHCM), Istituto Italiano di Tecnologia, Genova, Italy

*Address all correspondence to: luca.muratore@iit.it

IntechOpen

© 2021 The Author(s). Licensee IntechOpen. This chapter is distributed under the terms of the Creative Commons Attribution License (http://creativecommons.org/licenses/by/3.0), which permits unrestricted use, distribution, and reproduction in any medium, provided the original work is properly cited. CC BY

References

[1] A. Elkady and T. Sobh, "Robotics middleware: A comprehensive literature survey and Attribute-Based bibliography," *Journal of Robotics,* 2012. [Online]. Available: http://dx.doi.org/10.1155/2012/959013

[2] G. C. Buttazzo, *Hard Real-time Computing Systems: Predictable Scheduling Algorithms And Applications (Real-Time Systems Series)*. Santa Clara, CA, USA: Springer-Verlag TELOS, 2004.

[3] M. Aragão, P. Moreno, and *A. Bernardino*, "Middleware interoperability for robotics: A ros–yarp framework," *Frontiers in Robotics and AI*, vol. 3, p. 64, 2016. [Online]. Available: https://www.frontiersin.org/article/10.3389/frobt.2016.00064

[4] H. Bruyninckx, "OROCOS: design and implementation of a robot control software framework," *Proc. IEEE RAS EMBS Int. Conf. Biomed. Robot. Biomechatron.*, 2002. [Online]. Available: https://pdfs.semanticscholar.org/f32c/9806be8bd1a702a9732fc9cbe1626b3d37e6.pdf

[5] N. Ando, T. Suehiro, K. Kitagaki, T. Kotoku, and W.-K. Yoon, "Rt-middleware: distributed component middleware for rt (robot technology)," in *2005 IEEE/RSJ International Conference on Intelligent Robots and Systems*. IEEE, 2005, pp. 3933–3938. [Online]. Available: https://doi.org/10.1109/IROS.2005.1545521

[6] G. Metta, P. Fitzpatrick, and L. Natale, "Yarp: Yet another robot platform," *International Journal on Advanced Robotics Systems*, 2006. [Online]. Available: http://journals.sagepub.com/doi/pdf/10.5772/5761

[7] M. Quigley, K. Conley, B. P. Gerkey, J. Faust, T. Foote, J. Leibs, R. Wheeler, and A. Y. Ng, "Ros: an open-source robot operating system," in *ICRA Workshop on Open Source Software*, 2009.

[8] L. Jeongsoo, L. Jungho, and O. Jun-Ho, "Development of robot software framework podo: Toward multi-processes and multi-users," *Workshop on software architectures and methodologies for developing humanoid robots, IEEE HUMANOIDS 2014*, 2014. [Online]. Available: http://blog.pal-robotics.com/wp-content/uploads/2014/09/Lim_WSAH.pdf

[9] T. Houliston, J. Fountain, Y. Lin, A. Mendes, and others, "NUClear: A loosely coupled software architecture for humanoid robot systems," *Frontiers in Robotics*, 2016. [Online]. Available: https://doi.org/10.3389/frobt.2016.00020

[10] J. Smith, D. Stephen, A. Lesman, and J. Pratt, "Real-time control of humanoid robots using openjdk," in *Proceedings of the 12th International Workshop on Java Technologies for Real-time and Embedded Systems*, ser. JTRES '14. New York, NY, USA: ACM, 2014, pp. 29:29–29:36. [Online]. Available: http://doi.acm.org/10.1145/2661020.2661027

[11] M. Johnson, B. Shrewsbury, S. Bertrand, T. Wu, D. Duran, M. Floyd, P. Abeles, D. Stephen, N. Mertins, A. Lesman, J. Carff, W. Rifenburgh, P. Kaveti, W. Straatman, J. Smith, M. Griffioen, B. Layton, T. de Boer, T. Koolen, P. Neuhaus, and J. Pratt, "Team IHMC's lessons learned from the DARPA robotics challenge trials," *J. Field Robotics*, vol. 32, no. 2, pp. 192–208, 2015. [Online]. Available: http://onlinelibrary.wiley.com/doi/10.1002/rob.21571/abstract

[12] L. Muratore, A. Laurenzi, E. M. Hoffman, A. Rocchi, D. G. Caldwell, and N. G. Tsagarakis, "XBotCore: A Real-Time Cross-Robot Software

Platform," in *IEEE International Conference on Robotic Computing*, 2017. [Online]. Available: https://doi.org/10.1109/IRC.2017.45

[13] L. Muratore, A. Laurenzi, E. Hoffman, A. Rocchi, D. Caldwell, and N. Tsagarakis, "On the design and evaluation of xbotcore, a cross-robot real-time software framework," *Journal of Software Engineering for Robotics*, Jun. 2017. [Online]. Available: https://joser.unibg.it/index.php?journal=joser&page=article&op=viewFile&path[]=112&path[]=46

[14] L. Muratore, A. Laurenzi, E. M. Hoffman, and N. G. Tsagarakis, "The xbot real-time software framework for robotics: From the developer to the user perspective," *IEEE Robot. Autom. Mag.*, vol. 27, no. 3, pp. 133–143, 2020.

[15] P. Gerum, "Xenomai-implementing a rtos emulation framework on gnu/linux," *White Paper, Xenomai*, p. 81, 2004.

[16] J. H. Brown and B. Martin, "How fast is fast enough? choosing between xenomai and linux for real-time applications," *Twelfth Real-Time Linux Workshop*, 2012. [Online]. Available: https://pdfs.semanticscholar.org/9eb5/1dbe38fb23034e80b8664d8281996d2a5ef6.pdf?_ga=2.115305735.1422742923.1510677552-1828769110.1510677552

[17] A. Barbalace, A. Luchetta, G. Manduchi, *M. Moro*, A. Soppelsa, and C. Taliercio, "Performance comparison of vxworks, linux, rtai, and xenomai in a hard real-time application," *IEEE Transactions on Nuclear Science*, vol. 55, no. 1, pp. 435–439, 2008.

[18] G. F. Rigano, L. Muratore, A. Laurenzi, E. M. Hoffman, and N. G. Tsagarakis, "A mixed real-time robot hardware abstraction layer (r-hal)," *Encyclopedia with Semantic Computing and Robotic Intelligence*, vol. 02, no. 01, p. 1850010, 2018. [Online]. Available: https://doi.org/10.1142/S2529737618500107

[19] N. G. Tsagarakis, D. G. Caldwell, F. Negrello, W. Choi, L. Baccelliere, V. Loc, J. Noorden, L. Muratore, A. Margan, A. Cardellino *et al.*, "Walk-man: A high-performance humanoid platform for realistic environments," *Journal of Field Robotics*, vol. 34, no. 7, pp. 1225–1259, 2017.

[20] N. Kashiri, L. Baccelliere, L. Muratore, A. Laurenzi, Z. Ren, E. M. Hoffman, M. Kamedula, G. F. Rigano, J. Malzahn, S. Cordasco, P. Guria, A. Margan, and N. G. Tsagarakis, "Centauro: A hybrid locomotion and high power resilient manipulation platform," *IEEE Robotics and Automation Letters*, vol. 4, no. 2, pp. 1595–1602, 2019.

[21] N. G. Tsagarakis, S. Morfey, G. M. Cerda, L. Zhibin, and D. G. Caldwell, "Compliant humanoid coman: Optimal joint stiffness tuning for modal frequency control," in *Robotics and Automation (ICRA), 2013 IEEE International Conference on*. IEEE, 2013, pp. 673–678.

[22] N. G. Tsagarakis, D. G. Caldwell, F. Negrello, W. Choi, L. Baccelliere, V. Loc, J. Noorden, L. Muratore, A. Margan, A. Cardellino, L. Natale, E. Mingo Hoffman, H. Dallali, N. Kashiri, J. Malzahn, J. Lee, P. Kryczka, D. Kanoulas, M. Garabini, M. Catalano, M. Ferrati, V. Varricchio, L. Pallottino, C. Pavan, A. Bicchi, A. Settimi, A. Rocchi, and A. Ajoudani, "WALK-MAN: A High-Performance Humanoid Platform for Realistic Environments," *Journal of Field Robotics,* Jun. 2017. [Online]. Available: http://doi.wiley.com/10.1002/rob.21702

[23] N. G. Tsagarakis, F. Negrello, M. Garabini, W. Choi, L. Baccelliere, V. G. Loc, J. Noorden, M. Catalano, M. Ferrati, L. Muratore, P. Kryczka, E. M.

Hoffman, A. Settimi, A. Rocchi, A. Margan, S. Cordasco, D. Kanoulas, A. Cardellino, L. Natale, H. Dallali, J. Malzahn, N. Kashiri, V. Varricchio, L. Pallottino, C. Pavan, J. Lee, A. Ajoudani, D. G. Caldwell, and A. Bicchi, *WALK-MAN Humanoid Platform*. Cham: Springer International Publishing, 2018, pp. 495–548. [Online]. Available: https://doi.org/10.1007/978-3-319-74666-1_13

[24] M. Ferrati, A. Settimi, L. Muratore, A. Cardellino, A. Rocchi, E. Mingo Hoffman, C. Pavan, D. Kanoulas, N. G. Tsagarakis, L. Natale, and L. Pallottino, "The walk-man robot software architecture," *Frontiers in Robotics and AI*, vol. 3, p. 25, 2016. [Online]. Available: https://www.frontiersin.org/article/10.3389/frobt.2016.00025

[25] E. Mingo Hoffman, A. Rocchi, A. Laurenzi, and N. G. Tsagarakis, "Robot control for dummies: Insights and examples using opensot," in *17th IEEE-RAS International Conference on Humanoid Robots, Humanoids 2017, Birmingham, UK, November 15–17, 2017*, 2017.

[26] N. Vahrenkamp, M. Wächter, M. Kröhnert, K. Welke, and T. Asfour, "The robot software framework ArmarX," *it - Information Technology*, vol. 57, no. 2, 2015. [Online]. Available: https://doi.org/10.1515/itit-2014-1066

[27] P. Kaiser, D. Kanoulas, M. Grotz, L. Muratore, A. Rocchi, E. M. Hoffman, N. G. Tsagarakis, and T. Asfour, "An affordance-based pilot interface for high-level control of humanoid robots in supervised autonomy," in *IEEE/RAS International Conference on Humanoid Robots (Humanoids)*, 2016. [Online]. Available: https://doi.org/10.1109/HUMANOIDS.2016.7803339

[28] A. Paikan, D. Schiebener, M. Wächter, T. Asfour, G. Metta, and L. Natale, "Transferring object grasping knowledge and skill across different robotic platforms," in *Advanced Robotics (ICAR), 2015 International Conference on*, Jul. 2015, pp. 498–503.

[29] L. Baccelliere, N. Kashiri, L. Muratore, A. Laurenzi, M. Kamedula, A. Margan, J. Malzahn, and N. G. Tsagarakis, "Development of a human size and strength compatible bi-manual platform for realistic heavy manipulation tasks," in *2017 IEEE/RSJ International Conference on Intelligent Robots and Systems (IROS 2017)*, 2017. [Online]. Available: https://doi.org/10.1109/IROS.2017.8206447

[30] T. Klamt, M. Schwarz, C. Lenz, L. Baccelliere, D. Buongiorno, T. Cichon, A. DiGuardo, D. Droeschel, M. Gabardi, M. Kamedula, N. Kashiri, A. Laurenzi, D. Leonardis, L. Muratore, D. Pavlichenko, A. S. Periyasamy, D. Rodriguez, M. Solazzi, A. Frisoli, M. Gustmann, J. Rossmann, U. Suss, N. G. Tsagarakis, and S. Behnke, "Remote mobile manipulation with the centauro robot: Full-body telepresence and autonomous operator assistance," *Journal of Field Robotics*. [Online]. Available: https://onlinelibrary.wiley.com/doi/abs/10.1002/rob.21895

Chapter 2

Use Improved Differential Evolution Algorithms to Handle the Inverse Kinetics Problem for Robots with Residual Degrees of Freedom

Trung Nguyen and Tam Bui

Abstract

In this study, the Self-adaptive strategy algorithm for controlling parameters in Differential Evolution algorithm (ISADE) improved from the Differential Evolution (DE) algorithm, as well as the upgraded version of the algorithms has been applied to solve the Inverse Kinetics (IK) problem for the redundant robot with 7 Degree of Freedom (DoF). The results were compared with 4 other algorithms of DE and Particle Swarm Optimization (PSO) as well as Pro-DE and Pro-PSO algorithms. These algorithms are tested in three different Scenarios for the motion trajectory of the end effector of in the workspace. In the first scenario, the IK results for a single point were obtained. 100 points randomly generated in the robot's workspace was input parameters for Scenario 2, while Scenario 3 used 100 points located on a spline in the robot workspace. The algorithms were compared with each other based on the following criteria: execution time, endpoint distance error, number of generations required and especially quality of the joints' variable found. The comparison results showed 2 main points: firstly, the ISADE algorithm gave much better results than the other DE and PSO algorithms based on the criteria of execution time, endpoint accuracy and generation number required. The second point is that when applying Pro-ISADE, Pro-DE and Pro-PSO algorithms, in addition to the ability to significantly improve the above parameters compared to the ISADE, DE and PSO algorithms, it also ensures the quality of solved joints' values.

Keywords: differential evolution (DE), particle swarm optimization (PSO), inverse kinematic (IK), degree of freedom (DOF), optimization

1. Introduction

The robot Inverse Kinematics problem involves finding the joints' variable values that match input parameters of position and direction of the end effector [1]. These matched variable values will ensure that subsequent robot control will follow the desired trajectory. This is one of the important issues in the robotic field because it is related to other aspects such as motion planning, dynamic analysis and control [2]. Traditionally, there are several methods to resolve inverse kinematics problem

IntechOpen

for robots such as: geometry method is the method using geometric and trigonometric relationships to solve; the iterative method is often required inversion of a Jacobian matrix, etc. However, when applying these methods to solve the IK problem for robots, especially with redundant robots, it is often much more complicated and time-consuming. The reason is the nonlinearity of the formulas and the geometry between the workspace and the joint space. In addition, the difficult point is in the singularity, the multiple solutions of these formulas as well as the necessary variation of the formulas corresponding to the changes of different robot structures [3–5].

In addition to those existing methods of solving the IK problems, in recent years, the application of meta-heuristic optimization algorithms has become increasingly common. 8 optimization algorithms applied in [5] in the cases of a single point or a whole trajectory endpoint. The simulation results showed that the PSO algorithm can effectively solve the IK problem. In [6] the authors used algorithms such as ABC, PSO, and FA to solve the inverse kinematic requirement for Kawasaki RS06L 6-DoF robot in the task of picking and place objects. Ayyıldız et al. compared the results of all IK tests for a 4-DOF serial robot using 4 different algorithms: PSO, QPSO, GA and GSA [7]. Two versions of the PSO algorithm have been used to solve the IK problem for robots with a number of degrees of freedom from 9 to 180 [8]. In recent research [9], Malek et al. used PSO algorithm to handle inverse kinematics for a 7-DoF robot arm manipulator. The study mentioned both the requirements for the location and the direction of the endpoint, however, it only solved for 2 different end effector positions. Laura et al., in [10] used DE algorithm for the IK problem of 7-DoF robot. The problem was solved for specific points, but the quality evaluation parameters such as endpoint position deviation, execution time as well as the values of the joints' variable did not reach impressive quality. Ahmed El-Sherbiny et al. [11] proposed to use ABC variant algorithm for solving inverse kinematics problem in 5 DoFs robot arm. Serkan Dereli et al. [12] used a quantum behave partial algorithm (QPSO) for a 7-DoF serial manipulator and compare the results with other techniques such as firefly algorithm (FA), PSO and ABC.

In this study, the self-adaptive control parameters in Differential Evolution (ISADE) algorithm, that developed [13, 14] by authors, was applied to solve the problem of inverse kinematic for a 7-DOF serial robot. To compare the results, this IK problem was also handled by applying DE and PSO algorithms. In addition, the study also compared the results in the application of the above algorithms with the search space improvement of joints' variables (Pro-ISADE, Pro-PSO and Pro-DE) [15].

The remainder of the paper is divided into the following sections: Section II describes the experimental model. The theory of the PSO, DE and ISADE algorithms as well as the algorithms with improved search area, Pro-PSO, Pro-DE and Pro-ISADE, will then be presented in Section III. Section IV covers scenarios and object functions that will be applied to calculate the IK. The results after applying the algorithm are shown and compared in Section V. Finally, the conclusions are outlined in Section VI.

2. Testing model

The residual driven robots have many advantages such as easy escape from obstacles, flexible movement as well as a large operation space. However, their disadvantage is the complexity of the robot structure [16]. In this study, a serial redundant manipulator robot was used to evaluate the algorithm in resolving the inverse kinematics requirements. The simplified robot model was shown in the **Figure 1**. As in the figure, this serial robot manipulator is of type 7R (R: Revolute). The parameters of the D-H table of the robot are given in **Table 1**.

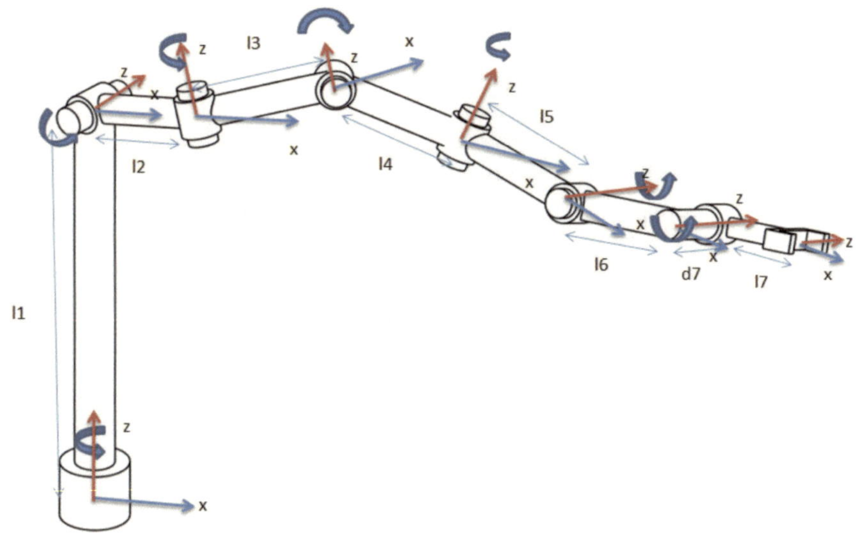

Figure 1.
The 7-DFO robot Scheme and coordinate systems used in the study.

Joint	θ(rad)	d(mm)	a(mm)	α(rad)
1	$-\pi < q_1 < \pi$	$d_1 = 500$	0	$-\pi/2$
2	$-\pi/2 < q_2 < \pi/6$	0	$l_2 = 200$	$\pi/2$
3	$-\pi/2 < q_3 < 2\pi/3$	0	$l_3 = 250$	$-\pi/2$
4	$-\pi/2 < q_4 < \pi/2$	0	$l_4 = 300$	$\pi/2$
5	$-\pi/2 < q_5 < \pi/2$	0	$l_5 = 200$	$-\pi/2$
6	$-\pi/2 < q_6 < \pi/2$	0	$l_6 = 200$	0
7	$-\pi/2 < q_7 < \pi/2$	$d_7 = 5$	$l_7 = 100$	0

Table 1.
D-H parameters.

The homogeneous transformation matrix can be used to obtain the forward kinematics of the robot manipulator, using the DH parameters in Eq. (1) [17].

$$T_{i-1i} = \begin{bmatrix} C\theta_i & -S\theta_i & 0 & a_i \\ S\theta_i C\alpha_i & C\theta_i C\alpha_i & -S\alpha_i & -d_i S\alpha_i \\ S\theta_i S\alpha_i & S\theta_i S\alpha_i & -C\alpha_i & -d_i C\alpha_i \\ 0 & 0 & 0 & 1 \end{bmatrix} \tag{1}$$

where S and C denote the sine and cosine functions.
The position and orientation of the end-effector can be determined by Eq. (2):

$$T_{07} = T_{01} * T_{12} * T_{23} * T_{34} * T_{45} * T_{56} * T_{67} = \begin{bmatrix} n_x & s_x & a_x & x_5 \\ n_y & s_y & a_y & y_5 \\ n_z & s_z & a_z & z_5 \\ 0 & 0 & 0 & 1 \end{bmatrix} \tag{2}$$

With:

$$T_{01} = \begin{bmatrix} cq1 & 0 & -sq1 & 0 \\ sq1 & 0 & cq1 & 0 \\ 0 & -1 & 0 & l1 \\ 0 & 0 & 0 & 1 \end{bmatrix} \tag{3}$$

$$T_{12} = \begin{bmatrix} cq2 & 0 & sq2 & l2cq2 \\ sq2 & 0 & -cq2 & l2sq2 \\ 0 & 1 & 0 & 0 \\ 0 & 0 & 0 & 1 \end{bmatrix} \tag{4}$$

$$T_{23} = \begin{bmatrix} cq3 & 0 & -sq3 & l3cq3 \\ sq3 & 0 & cq3 & l3sq3 \\ 0 & -1 & 0 & 0 \\ 0 & 0 & 0 & 1 \end{bmatrix} \tag{5}$$

$$T_{34} = \begin{bmatrix} cq4 & 0 & sq4 & l4cq4 \\ sq4 & 0 & -cq4 & l4sq4 \\ 0 & 1 & 0 & 0 \\ 0 & 0 & 0 & 1 \end{bmatrix} \tag{6}$$

$$T_{45} = \begin{bmatrix} cq5 & 0 & -sq5 & l5cq5 \\ sq5 & 0 & cq5 & l5sq5 \\ 0 & -1 & 0 & 0 \\ 0 & 0 & 0 & 1 \end{bmatrix} \tag{7}$$

$$T_{56} = \begin{bmatrix} cq6 & -sq6 & 0 & l6cq6 \\ sq6 & cq6 & 0 & l6sq6 \\ 0 & 0 & 1 & 0 \\ 0 & 0 & 0 & 1 \end{bmatrix} \tag{8}$$

$$T_{67} = \begin{bmatrix} cq7 & -sq7 & 0 & l7cq7 \\ sq7 & cq7 & 0 & l7sq7 \\ 0 & 0 & 1 & d7 \\ 0 & 0 & 0 & 1 \end{bmatrix} \tag{9}$$

Where, T_{07} is matrix to produce a Catesian coordinate for any seven joint values. In the Eq. (10), (x_E, y_E, z_E) denote the elements of position vector whereas, $(n_x, n_y, n_z, s_x, s_y, s_z, a_x, a_y, a_z)$ are the rotational elements of transformation matrix. In this study, only position vectors were used to calculate the distance error. After the computation, the end-effector coordinate in the manipulation space is determined by:

$$
\begin{aligned}
x_E = {} & d7sq5cq4sq1sq3 - cq1cq2cq3 + cq1sq2sq4 - cq5cq3sq1 + cq1cq2sq3 - l4cq4sq1sq3 \\
& - cq1cq2cq3 - l5sq5cq3sq1 + cq1cq2sq3 + l2cq1cq2 - l3sq1sq3 - l5cq5cq4sq1sq3 \\
& - cq1cq2cq3 + cq1sq2sq4 + l7cq7sq6sq4sq1sq3 - cq1cq2cq3 - cq1cq4sq2 \\
& - cq6cq5cq4sq1sq3 - cq1cq2cq3 + cq1sq2sq4 + sq5cq3sq1 + cq1cq2sq3 \\
& + l6sq6sq4sq1sq3 - cq1cq2cq3 - cq1cq4sq2 + l7sq7cq6sq4sq1sq3 - cq1cq2cq3 \\
& - cq1cq4sq2 + sq6cq5cq4sq1sq3 - cq1cq2cq3 + cq1sq2sq4 + sq5cq3sq1 \\
& + cq1cq2sq3 - l6cq6cq5cq4sq1sq3 - cq1cq2cq3 + cq1sq2sq4 + sq5cq3sq1 \\
& + cq1cq2sq3 + l3cq1cq2cq3 - l4cq1sq2sq4 \\
y_E = {} & l4cq4cq1sq3 + cq2cq3sq1 - d7sq5cq4cq1sq3 + cq2cq3sq1 - sq1sq2sq4 - cq5cq1cq3 \\
& - cq2sq1sq3 + l5sq5cq1cq3 - cq2sq1sq3 + l2cq2sq1 + l3cq1sq3 + l5cq5cq4cq1sq3 \\
& + cq2cq3sq1 - sq1sq2sq4 - l6sq6sq4cq1sq3 + cq2cq3sq1 + cq4sq1sq2 \\
& - l7cq7sq6sq4cq1sq3 + cq2cq3sq1 + cq4sq1sq2 - cq6cq5cq4cq1sq3 + cq2cq3sq1 \\
& - sq1sq2sq4 + sq5cq1cq3 - cq2sq1sq3 - l7sq7cq6sq4cq1sq3 + cq2cq3sq1 \\
& + cq4sq1sq2 + sq6cq5cq4cq1sq3 + cq2cq3sq1 - sq1sq2sq4 + sq5cq1cq3 - cq2sq1sq3 \\
& + l6cq6cq5cq4cq1sq3 + cq2cq3sq1 - sq1sq2sq4 + sq5cq1cq3 - cq2sq1sq3 \\
& + l3cq2cq3sq1 - l4sq1sq2sq4 \\
z_E = {} & l1 - l2sq2 + d7sq5cq2sq4 + cq3cq4sq2 + cq5sq2sq3 - l5cq5cq2sq4 + cq3cq4sq2 \\
& - l6sq6cq2cq4 - cq3sq2sq4 - l3cq3sq2 - l4cq2sq4 - l6cq6cq5cq2sq4 + cq3cq4sq2 \\
& - sq2sq3sq5 - l7cq7cq6cq5cq2sq4 + cq3cq4sq2 - sq2sq3sq5 + sq6cq2cq4 \\
& - cq3sq2sq4 + l7sq7sq6cq5cq2sq4 + cq3cq4sq2 - sq2sq3sq5 - cq6cq2cq4 \\
& - cq3sq2sq4 - l4cq3cq4sq2 + l5sq2sq3sq5
\end{aligned}
\tag{10}
$$

When solving the problem of inverse kinematics, with the endpoint coordinates as on the left side of Eq. (10), we need to find the values of the matching variable q. However, according to the Equation, the number of equations is much less than the number of variables. This makes it very difficult to find a unique and exact matching solution. In this study, ISADE algorithm and ISADE with searching space improvement algorithm (Pro-ISADE) were used to solve the IK problem for the robot. To compare the results, the study also used some other optimization algorithms such as PSO, DE as well as Pro-PSO and Pro-DE to solve the same IK problem for the robot above.

3. Applied algorithms and object functions

3.1 PSO

Particle swarm optimization was developed flying Kenney and Eberhart [18, 19] based on observing the moving characteristics of bird flock and fish school. In this algorithm the individual of the population is called particle. The particle of the population (Called swarm) can move in its space and offer a potential solution. Particles can memorize best condition and find and exchange information to other members. Each particle in the population has two characteristics: position and velocity. Starting with the particle population, each particle monitors its coordinates and updates position and speed according to the best solution for each iteration. The velocity and position values are shown in the following equation:

$$
\begin{aligned}
& v_{id}(t+1) = \mathrm{w}v_{id}(t) + c_1 rand\left[p_{id}(t) - x_{id}(t)\right] + c_2 rand\left[g_{id}(\mathrm{t}) - \mathrm{x}_i(\mathrm{t})\right] \\
& \mathrm{x}_{id}(t+1) = x_{id}(t) + v_{it}(t)
\end{aligned}
\tag{11}
$$

In particular, x_{xi}, v_i are the position and velocity of the particle i-th, respectively; d is number of dimension; w is the inertia weight factor, $c_1$and $c_2$are cognitive learning rate and social learning rate, respectively; pi is the *pbest* value of *i_th* particle; What is *gbest* value of the population.

3.2 DE

Differential Evolution (DE) algorithm is a population-based stochastic optimization algorithm recently introduced. DE works with two populations; old generation and new generation of the same population. The population is randomly initialized within the initial parameter bounds individuals in the population has two characteristics: position and velocity. Starting with the individual population, each individual monitors its coordinates and updates position and speed according to the best solution for each iteration. Velocity values (V) is randomly created in one of eight ways:

$$\begin{aligned}
&V = X_{r1} + F_w(X_{r2} - X_{r3}) \\
&V = X_{best} + F_w(X_{r1} - X_{r2}) \\
&V = X_{r1} + F_w(X_{r2} - X_{r3}) + F_w(X_{r4} - X_{r5}) \\
&V = X_{best} + F_w(X_{r1} - X_{r2}) + F_w(X_{r3} - X_{r4}) \\
&V = X + F_w(X_{r1} - X_{r2}) + F_w(X_{best} - X) \\
&V = X + F_w(X_{r1} - X_{r2}) + F_w(X_{r3} - X_{r4}) + F_w(X_{best} - X) \\
&V = X_{r1} + F_w(X_{r2} - X_{r3}) + F_w(X_{best} - X_{r1}) \\
&V = X + F_w(X_{r2} - X_{r3}) + F_w(X_{r1} - X)
\end{aligned} \quad (12)$$

In particular, F is Scaling factor, r_1, r_2, r_3, r_4, r_5 is random solution, $r_1, r_2, r_3, r_4, r_5 \in \{1, 2, 3, \ldots, N_p\}$ and $r_1 \neq r_2 \neq r_3 \neq r_4 \neq r_5 \neq$ i, X_{best} is population filled with the best member.

Position values new (U) shown in the following Equation:

$$U = X. * FM_{mpo} + V. * FM_{mui} \quad (13)$$

FM_{mui} are all random numbers <0.9, FM_{mpo} is inverse mask to $FM_{mui.}$

3.3 ISADE

In the [13, 14], we suggested to develop a new version of DE algorithm that can automatically adapt the learning strategies and the parameters settings during evolution. The main ideas of the ISADE algorithm are summarized below.

3.3.1 Mutation operator

ISADE probabilistically selects one out of several available learning strategies in the mutation operator for each individual in the current population. In this research, we select three learning strategies in the mutation operator as candidates: “DE/best/1/bin”, “DE/best/2/bin” and “DE/rand to best/1/bin” that are respectively expressed as:

$$DE/best/1: \ V_{i,j}^G = X_{best,j}^G + F * \left(X_{r1,j}^G - X_{r2,j}^G\right) \quad (14)$$

$$DE/best/2 : V_{i,j}^{G} = X_{best,j}^{G} + F * \left(X_{r1,j}^{G} - X_{r2,j}^{G}\right) + F * \left(X_{r3,j}^{G} - X_{r4,j}^{G}\right) \quad (15)$$

$$DE/rand\,to\,best/1 : V_{i,j}^{G} = X_{r1,j}^{G} + F * \left(X_{best,j}^{G} - X_{r1,j}^{G}\right) + F * \left(X_{r2,j}^{G} - X_{r3,j}^{G}\right) \quad (16)$$

Where: $i = \{1, 2, \dots, NP\}; j = \{1, \dots, D\}$ are current population and design variable, respectively.

$"DE/Randtobest/1/bin"$ strategy usually demonstrates good diversity while the $"DE/best/1/bin"$ and $"DE/best/2/bin"$ strategy show good convergence property, which we also observe in our trial experiments.

3.3.2 Adaptive scaling factor F and crossover control parameter CR

In the ISADE algorithm, the author suggested to use the sigmoid function to control neighborhood parameter. we sort the particles by estimating their fitness. A ranked particle is labeled with ranked number and assigned F that corresponds with its number. The formula for F by sigmoid function as following:

$$F_i = \frac{1}{1 + \exp\left(\alpha * \frac{i - \frac{NP}{2}}{NP}\right)} \quad (17)$$

Where: α, idenote the gain of the sigmoid function, particle of the i_{th}in NP, respectively.

For better performance of ISADE, the scale factor F should be high in the beginning to have much exploration and after curtain generation F needs to be small for proper exploitation. Thus, we proposed to calculate the F as follow:

$$F_{iter}^{mean} = F_{\min} + (F_{\max} - F_{\min})\left(\frac{iter_{\max} - iter}{iter_{\max}}\right)^{n_{iter}} \quad (18)$$

Where: $F_{max}, F_{min}, iter, iter_{max}$ *and* n_{iter} are the lower boundary condition of F, upper boundary condition of F, current generation, maximum generation and nonlinear modulation index, respectively.

The author introduced a novel approach of scale factorF_i of each particle with their fitness in Eq. (15). Thus, in one generation the value of $F_i^{iter}(i = 1, \dots, NP)$ are not the same for all particles in the population rather they are changed in each generation. The final value of scale factor for each generation is calculated as follow:

$$F_{iter}^{i} = \frac{F_i - F_{iter}^{mean}}{2} \quad (19)$$

Where $iter = 1, \dots, iter_{max}$ *and* $i = 1, \dots, NP$

The control parameter CR is adapted as following:

$$CR_i^{G+1} = \begin{cases} rand_2\, if\ rand_1 \leq \tau \\ CR_i^{G}\ othewise \end{cases} \quad (20)$$

The ISADE algorithm was summarized as in the **Figure 2**.

3.4 Cost functions and Algorithms with searching space improvement

As mention in the introduction part, the disadvantage of many studies using optimization algorithms to solve the IK problem of redundant robots is to focus on

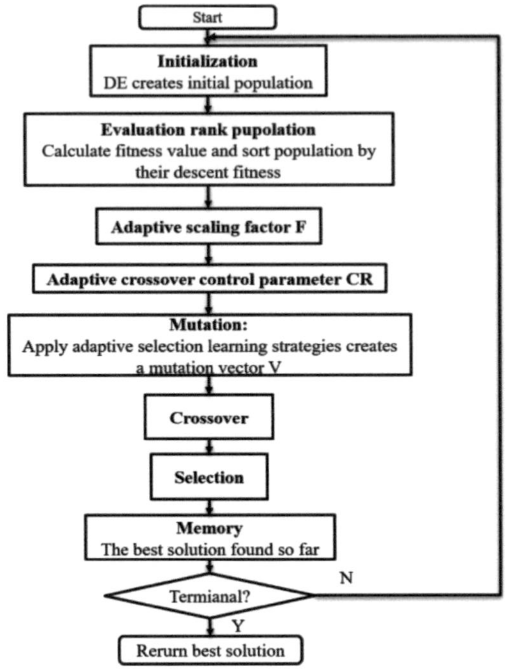

Figure 2.
ISADE Flowchart.

the results related to the optimal running process such as execution time, number of generation ... but have not yet considered the feasibility of the joints' variable values. In order to overcome these drawbacks, the author of this research [15] proposed this algorithm that is explained as following: The solution to improve the continuity of joints' values constrains the initialization domain of X. This help the program to achieve the dual goal of increasing calculation speed, accuracy and ensuring continuity for the value of joints' variables. In this algorithm, firstly the robot from any position moves to the first point of the trajectory. With this first point, the initialization values for the particles are randomly selected in the full Range of Motion (RoM) of joints. In addition, the target function in this case has the form:

$$Func.1 = a * \sqrt{\sum_{k=1}^{5}(q_i^k - q_0^k)} + b * \sqrt{\begin{matrix}(x_i - x_{ei})^2 + (y_i - y_{ei})^2 + (z_i - z_{ei})^2 + (Rx_i - Rx_e)^2 + \\ (Rx - Rx_e)^2 + (Ry_i - Ry_e)^2 + (Rz_i - Rz_e)^2\end{matrix}} \quad (21)$$

where the values q_i^k *and* q_i^k (i = 1) are the joints' variable values at the original position and 1st point on the trajectory, respectively; (x_i, y_i, z_i) and (x_{ei}, y_{ei}, z_{ei}) are the End-effector coordinates for the i-point (i = 1) found by the algorithm and the desired End-effector coordinates; (Rx_i, Ry_i, Rz_i) and (Rx_{ei}, Ry_{ei}, Rz_{ei}) are corresponding rotation cosine angles performing orientation of the end-effector which are found by Algorithm and orientation of the desired end-effector; a, b are penalty coefficients. Cost function as Eq. (21) ensures the energy spent in the joints to reach the 1st desired position is minimized. Besides, it also minimizes the distance error between the actual and desired end-effector position. The condition to stop for points of trajectory is that the Cost Func.1 value is less than value of e or the number of iterations reaches 600 and the number of times algorithm running <10.

After calculating for 1st point of the trajectory, the remaining points are calculated with a search limitation around the previous optimal joints' values. By using this suggested range, the program's search space will be limited while ensuring the continuity of the joint variables. In this case, the target function is still the same as the function of 1st point, but it has coefficient a = 0.

4. Scenarios

4.1 Scenario 1

In Scenario 1, an endpoint in the workspace were randomly selected; the PSO; DE and ISADE algorithms were then applied to solve the required problem. The purpose of this Scenario is to compare the convergence speed of the three algorithms. In this case, since the initial and the desired endpoints can be far apart, the Pro-PSO; Pro-DE and Pro-ISADE algorithms cannot be applied.

4.2 Scenario 2

In this case, the robot was required to move the endpoint through 100 points in the robot's working space one after another. These points were selected at random for the purpose of testing the effectiveness of each algorithm with many distinct points. Similar to the previous case, in this Scenario we also only applied the algorithms PSO, DE and ISADE with the solution space of the matching variable which limits the motion of these joints.

4.3 Scenario 3

The manipulator robot was required to move the end effector following a certain trajectory. The selected trajectory is spiral, and it is described by the following function:

$$\begin{cases} x_E = 200 * cos\,(2 * z_E/100) \\ y_E = 200 * sin\,(2 * z_E/100) \\ z_E = n * pi \end{cases} \tag{22}$$

Where: (x_E, y_E, z_E) is the desired endpoint coordinate on the trajectory. With 6 algorithms of PSO, Pro ISO, DE, Pro-DE and ISADE, Pro-ISADE, the comparison of the results on the same graph is not favorable. Therefore, the study divided this case into two smaller Scenarios:

- Scenario 3.1: Results when using ISADE algorithm comparing with results from PSO and DE algorithms.
- Scenario 3.2: Compare the results using Pro-ISADE algorithm with the results getting from Pro-PSO and Pro-DE algorithms

And then results from Scenario 3.1 were be compared with the results from Scenario 3.2.

5. Simulation and results

5.1 Experimental setup

The main task of this study is to find the optimal value of the joints' variable to ensure the end effector of robots can reach the desired points. The desired point positions of the Scenario 2 and 3 are shown as the **Figure 3**. Research using the ISADE and Pro-ISADE algorithm, which were developed by the authors [13–15], to get simulation results of inverse kinematics problem and then compared it with the results when using PSO, DE and Pro-PSO, Pro-DE algorithms. When solving the IK problem for the 7-DoF serial robot manipulator, the study focused on three main aspects. The first of these is the sensitivity of the solution - in the other word, the amount distance error of end effector is minimum. The second criterion was the execution time. In order to avoid the endless loop, the maximum numbers of generation $(iter_{max})$were set as 600, 600 and 130 for PSO (Pro-PSO), DE (Pro-DE) and ISADE (Pro-ISADE), respectively. And the final aspect is the searching space of joints' variables. Normally, Normally, almost all studies have been using the Range of Motion (RoM) of joints for its boundary space. Our algorithm [15] proposed to use the searching space of current generation is around previous optimal joints' values. In the **Table 2**, the ubs_{i+1} *and* lbs_{i+1} are the joints' upper and lower boundary of the current generation.C_1 *and* C_2 are weights of personal best and global best, respectively. w is the inertia weight. ρ is the number of run for each algorithm to choose the best result. Besides, after some trial runs for the algorithms, we noticed that our ISDE algorithm gave much better results than DE and the least was the PSO algorithm. Thus, when setting up the maximum distance error by the fitness value setting for the end effector position, the study set the value of $1e-14\ (m)$; $1e-15\ (m)$ and $1e-17\ (m)$ for PSO (Pro-PSO); DE (Pro-DE) and ISADE (Pro-ISADE), respectively or that can be seen in the **Table 2**. In this research, the proposed and other methods were tested in the two different Scenarios. Both the first and second Scenario was coded by Matlab version 2019a and run on the computer equipped with an Intel Core i5-4258U @2.4GHz processor and 8 GB Ram memory.

5.2 Scenario 1 results

After applying the inverse kinematic problem processing algorithms for a single endpoint, the results are shown in **Figures 4** and **5**. All algorithms are able to handle

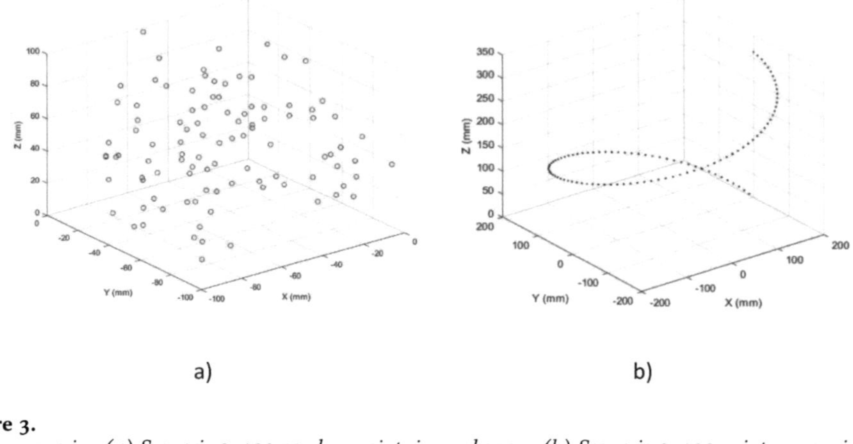

Figure 3.
Testing scenarios. (a) Scenario 2: 100 random points in workspace; (b) Scenario 3: 100 points on a spiral trajectory.

Algorithms	Max No. of Gen. max *iter*	Max Distance Err. (m)	Searching space $[ubs_{i+1}lbs_{i+1}]$	ρ	Mut. rate	Cross. rate	C_1	C_2	w	F_i
PSO	600	1e-14	*RoM*	10	**	**	1.5	1.5	$w = w_{start} + \left(\frac{iter}{iter_{max}}\right) * (w_{end} - w_{start})$	**
Pro-PSO	600	1e-14	$q_{oi} \pm \pi/100$		**	**				**
DE	600	1e-15	*RoM*		Scheme 2 in Eq. (10)	0.9	**	**	**	0.5
Pro-DE	600	1e-15	$q_{oi} \pm \pi/100$			0.9	**	**	**	0.5
ISADE	130	1e-17	*RoM*			*Eq*.(20)	**	**	**	*Eq*.(19)
Pro-ISADE	130	1e-17	$q_{oi} \pm \pi/100$				**	**	**	

Table 2.
Optimization parameters used in PSO, Pro-PSO, DE Pro-DE, and ISADE, Pro-ISADE.

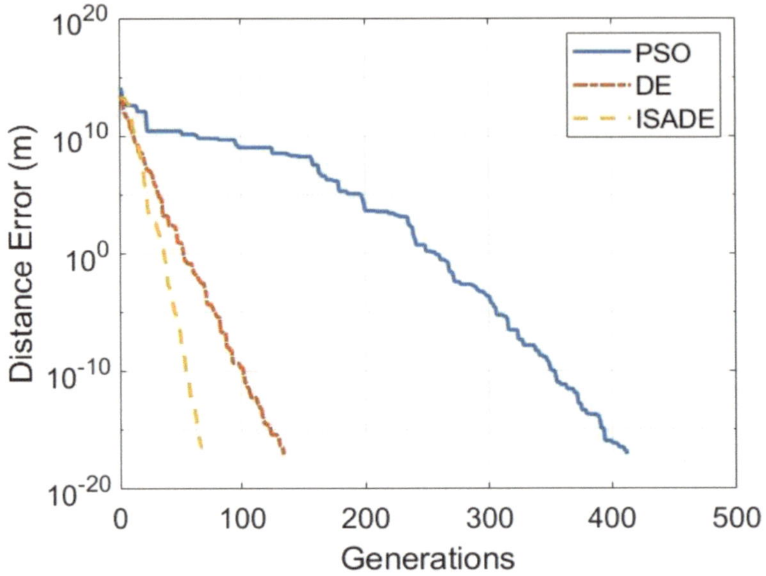

Figure 4.
End effector distance error vs. generations in Scenario 1.

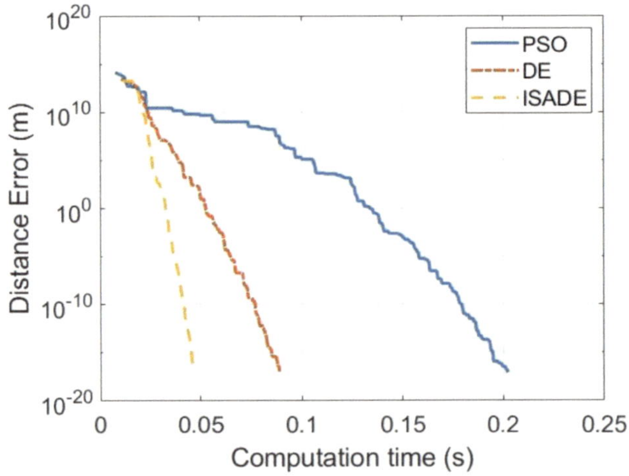

Figure 5.
End effector distance error vs. time in Scenario 1.

	Max. Iteration	Position error (m)	Calculation time (s)
PSO	85	2.6815e-04	0.0941
DE	85	5.7514e-10	0.0715
ISADE	85	2.8422e-13	0.0490

Table 3.
Comparison of ISADE with other algorithms.

the inverse kinetics problem, but the best results have been obtained with the ISADE algorithm as shown in **Table 3**.

Figures 4 and **5** show convergence speed of algorithms corresponding to the number of iterations and processing time, respectively. The results show that the processing speed of the ISADE algorithm is the best, followed by the DE algorithm and finally with the PSO algorithm. In **Table 3** the study of selecting stop conditions for algorithms is the maximum number of iterations of 85 rounds. After 10 runs, the best results are shown in the table. The ISADE algorithm gives the best processing results in terms of both quality and speed. The endpoint deviation can reach 2.8422e-13 (m) in 0.049 (s) time. For the PSO algorithm, it can handle the reverse kinematic problem for the end point with an accuracy of 2.6815e-4 in a period of 0.0941 (s). and, 5.7514e-10 (m) and 0.0715 (s) are the accuracy of end effector and execution time for DE algorithm.

5.3 Scenario 2 results

As mentioned above, in this Scenario 2, algorithms was used to resolve inverse kinematics problem for 100 randomly chosen points within the workspace of the robot. When processed at each point, the end effector started at the same initial position of [0 0 0 0 0 0 0] for 7 serial joints values. Because the end effector points all come from the same starting point to go to each of the 100 points, the study only used the ISADE algorithm and compares with the results from PSO and DE algorithms without using the Pro-ISADE algorithm as well as Pro-DE and Pro-PSO.

The 100 randomly selected points were shown in the **Figure 3a**. Results when applying ISADE and the other algorithm were presented in the **Figure 6**. As shown in the Figure, all algorithms have solved problem well. In particular, with the ISADE algorithm, although the fitness value in experimental setup required 1000 and 100 times higher than the required by applying the PSO and DE algorithms, respectively, it was not only guaranteed required precision but also showed faster processing speed and fewer iterations compared to the 2 other algorithms. Specifically, as shown in **Figure 6b** and **c** and especially **Table 4**, the average execution time when using ISADE to solve IK of each points was around 0.0685 second, while this value of the PSO and DE algorithm were on average 0.2307 (s) and 0.0978 (s) respectively. The main reason for this, as seen in **Figure 6b** and **Table 4**, was the number of generations to reach the optimal values much higher in PSO algorithm and slightly higher in DE algorithm, compared to in ISADE algorithm. Specifically, the PSO algorithm needed an average of 413.24 and the DE algorithm needed average of 124.45 loops to find a solution, while the ISADE algorithm used an average of 85.63 loops. Another remarkable thing is although there was not much difference in the number of iterations to solve the problem between the two algorithms DE and ISADE, but the ISADE algorithm still gave a processing speed of 1.42 times higher than DE algorithm though required 100 times more accuracy for the ISADE algorithm. This demonstrated the very high efficiency of the ISADE algorithm when it was applied to handle inverse kinematics problem for this robot. In short, in the optimization study for randomly chosen points in working space, the ISADE algorithm presented the best algorithm to resolve the IK requirement in term of accuracy, iteration and execution time.

5.4 Scenario 3 results

In Scenario 2, the end effector moved through the 100 points located on a specific trajectory that was defined in Eq. (22) and shown in **Figure 3b**. The main difference between Scenario 2 and Scenario 3 is that, instead of after solving each IK problem for each point, the end effector goes back to the original point to continue processing for the next points like in Scenario 2, in Scenario 3 the end effector starts from previous point in order to calculate for the next point. Stemming from this feature, the searching space of joints' variable also starts previous optimal joints' values. However, depending on the searching space we have 2 smaller cases such as:

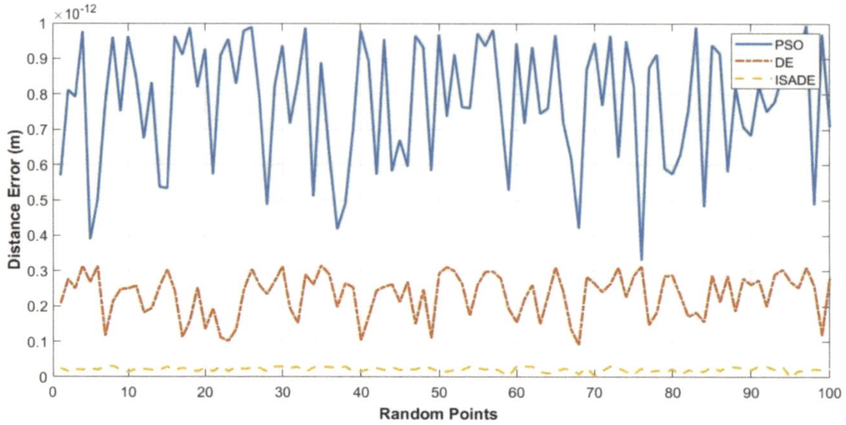

a)

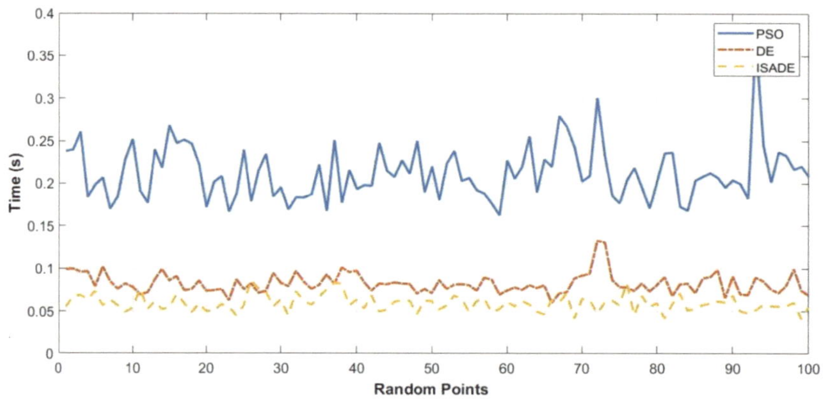

b)

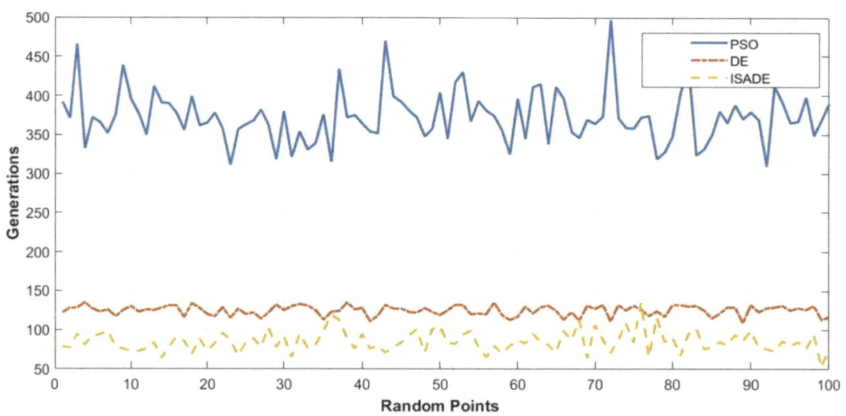

c)

Figure 6.
Results for Scenario 2. (a) Distance error. (b) Execution time. (c) Number of generations.

- Scenario 3.1: Searching spaces for joints' variables are *RoMs*. Then, like the Scenario 2, the study compared the results when using the ISADE algorithm with the results when using the PSO and DE algorithms.

	PSO	DE	ISADE
Fitness value	1e-14	1e-15	1e-17
Avg. error	7.3016e-13	2.2938e-13	2.1644e-14
STD	2.0415e-13	5.991e-14	6.2125e-15
Avg. iteration	413.24	124.45	85.63
Avg. execution time	0.2307	0.0978	0.0685

Table 4.
Comparative results in case 2.

- Scenario 3.2: Searching spaces for joints' variables are around the previous optimal joints' values. The study compared the results when using Pro-ISADE algorithm with when using Pro-PSO and Pro-DE algorithms.

The results were presented in the **Figure 7** and **Table 5**. Similar to the Scenario 2, although the experimental installation required the ISADE (and Pro-ISADE) algorithm to be 100 and 1000 times more accurate than the algorithm DE (Pro-DE) and PSO (Pro PSSO), respectively, all of 6 algorithms gave appropriated solutions for all the points in the trajectory. It can be seen that, in both cases 3.1 and 3.2 the ISADE and Pro-ISADE algorithms showed the best ability to resolve the inverse kinematics problems in all 3 aspects: accuracy, execution time and number of generations. More specifically, in Scenario 3.1, when searching space for joints' variables were *RoMs*, the average achieved accuracies for ISADE was around 2.0748e-14 (m) that is much better than the values of 7.5404e-13 (m) and 2.2260e-13 (m) corresponding for PSO and DE algorithms. Although the ISADE algorithm was set to a fitness value to achieve such higher accuracy, the execution time of the algorithm was still below the time of PSO and DE algorithm. These average execution time values were 0.0679 (s); 0.0845 (s) and 0.3478 (s) second for ISADE, DE and Pro algorithm, respectively. The above results can be partly explained based on the number of necessary iterations that each algorithm was needed to find the optimal values of joints variables. From **Figure 6c**, it showed that, when solving the IK problem for almost points in the spiral trajectory, the ISADE method used the least number of iterations. The **Table 5** presented more clearly, on the average each point in the trajectory the ISADE needed 85.19 generations to find the optimal values, these means number for DE and PSO algorithm are 125.44 and 391.1

In Scenario 3.2, the searching space for joints' variables were around previous optimal values that were set up as in the **Table 2**. Similar to the Scenario 3.1, all of the comparison parameters gotten from using Pro-ISADE algorithm were better than that values from Pro-DE and Pro-PSO algorithms. These parameters are described in the as well as **Table 5**. In order to comparison between Scenario 3.1 with Scenario 3.2, all average parameters was shown in the **Table 5**. From all comparison, the proposed ISADE or Pro-ISADE were always proved the best solution to solve the inverse kinematics requirements for the manipulator robot. Moreover, **Table 5** also showed that, the Pro-ISADE had better performance compared to ISADE. By using Pro-ISADE algorithm, it reduced all of parameters including distance error, execution time and number of generations.

Another very important result gotten from Scenario 3.2 is the quality of joints' values. **Figure 8** show the joints' value in two cases of using ISADE in Scenario 3.1 and using Pro-ISADE in Scenario 3.2. It is clear that the joints' value in the Scenario 3.1 were change dramatically. On the contrary, the values of joints in Scenario 3.2 changed continuously and slowly. The quality of joints variable values as **Figure 9b**,

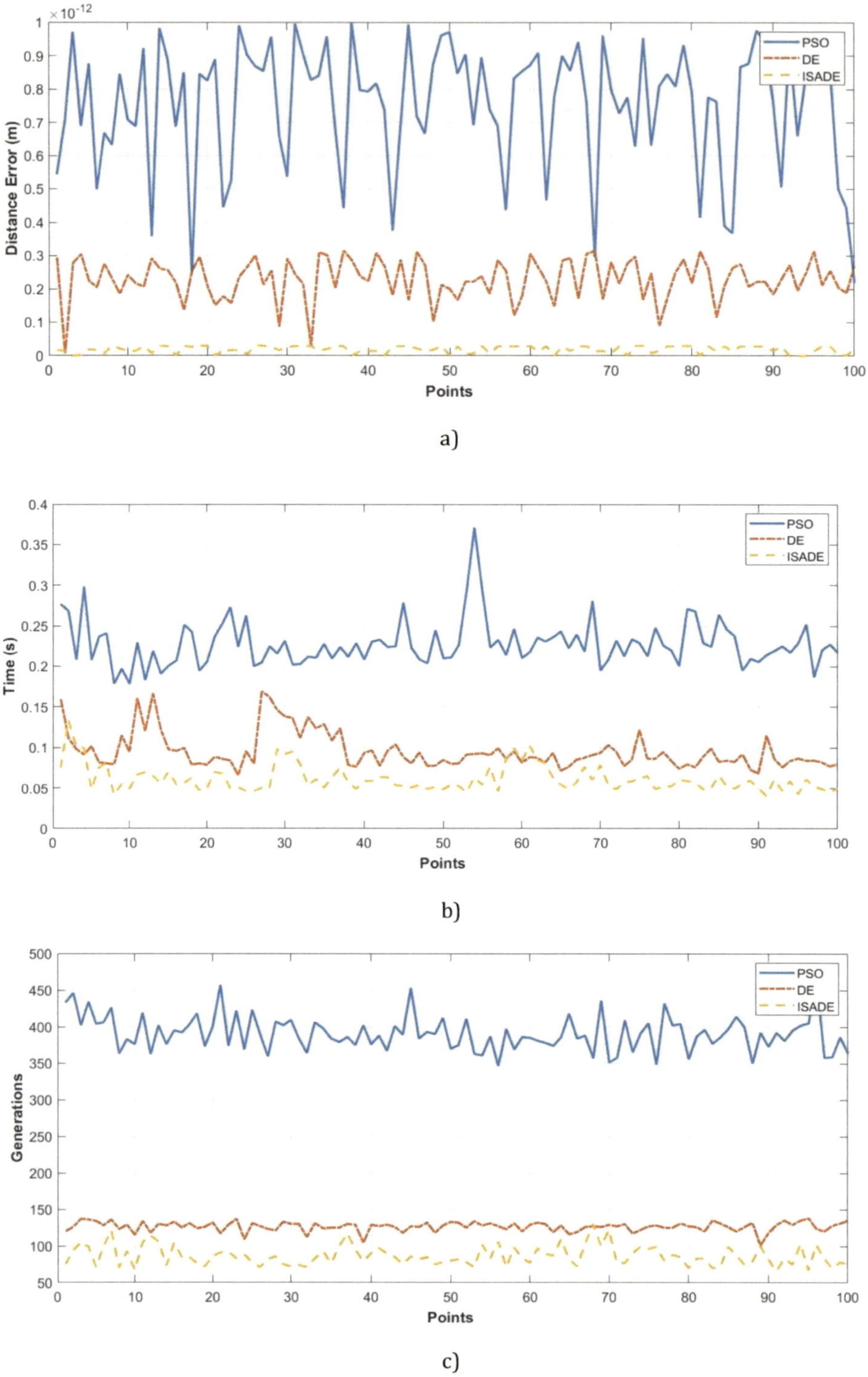

Figure 7.
Results for Scenario 3.1. (a) Distance error. (b) Execution time. (c) Number of generations.

that received by using Pro-ISADE, will ensure feasibility in the next stages of calculation and design for the robot. These values, along with the values of speed, acceleration, as well as the weight parameters of the stages, will be used in the dynamic problem as well as in future control.

	PSO	Pro-PSO	DE	Pro-DE	ISADE	Pro-ISADE
Scenario 1						
Fitness value	1e-9	Not applied	1e-10	Not applied	1e-12	Not applied
Avg. error (m)	2.4151e-09	Not applied	6.9655e-10	Not applied	6.8362e-11	Not applied
STD (m)	5.8117e-10	Not applied	2.0075e-10	Not applied	2.3796e-11	Not applied
Avg. iteration	357.91	Not applied	76.54	Not applied	64.34	Not applied
Avg. execution time (s)	0.2931	Not applied	0.1115	Not applied	0.0455	Not applied
Scenario 3.1 (Italic values) and Scenario 3.2						
Fitness value	*1e-14*	1e-14	*1e-15*		*1e-17*	1e-12
Avg. error (m)	*7.4140e-13*	7.4650e-13	*2.2260e-13*	2.2950e-13	*2.0748e-14*	2.0103e-14
STD (m)	*1.9574e-13*	1.9736e-13	*6.5615e-14*	6.1330e-14	*1.0414e-14*	9.8913e-15
Avg. iteration	*429.950*	407.8800	*125.4400*	114.2700	*85.1900*	75.2300
Avg. execution time (s)	*0.3604*	0.2576	*0.1015*	0.0845	*0.0679*	0.0554

Italics were used to differentiate the results of Scenario 3.1 and 3.2.

Table 5.
Comparative results between all cases.

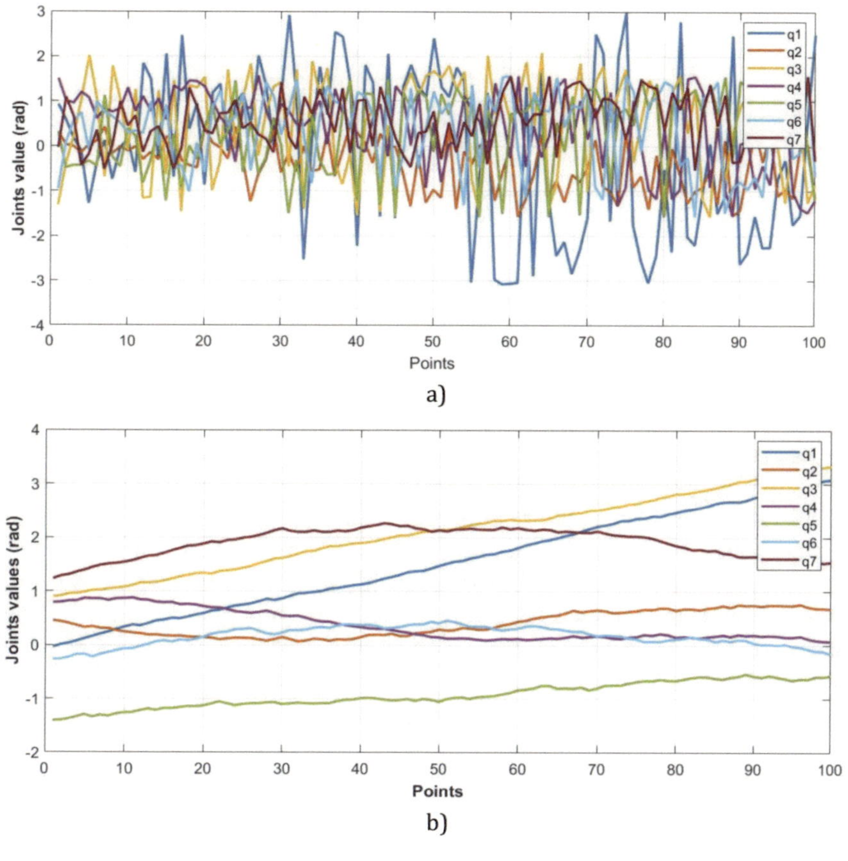

Figure 8.
Joint variables' results. (a) Joint variables' values in Scenario 3.1 using ISADE algorithm. (b) Joint variables' values in Scenario 3.2 using Pro-ISADE algorithm.

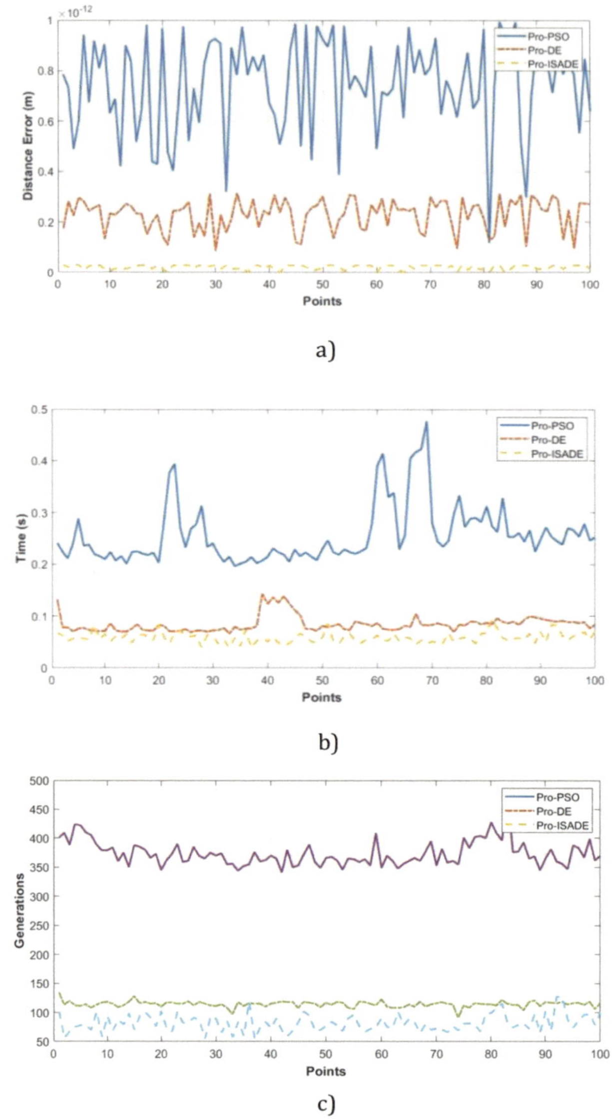

Figure 9.
Results for Scenario 3.2. (a) Distance error. (b) Execution time. (c) Number of generations.

In short, after comparing the results of Scenario 3.1 and 3.2, it is possible to conclude that the ISADE algorithm and Pro-ISADE are the best solutions to solve the IK problem for the robot in all aspects: endpoint accuracy, execution time and number of generation. The Pro-ISADE algorithm not only guarantees the above parameters, it also ensures the quality of the joints' variables to serve the next computational and design stages.

Table 5 summarizes results of the average error, the standard deviation of error (STD), the average iteration and the average execution time of all Scenarios. As in the table, the algorithms of ISADE and Pro-ISADE got the better results than the other algorithms.

As mentioned at the beginning of this article, intelligent optimization techniques have been using more and more popular in difficult and complex tasks including the IK problem for redundant manipulator robots. **Table 6** shows some studies used meta-heuristic optimization algorithms to resolve the inverse kinematics task for different robot models. The Table presents: the used algorithm for the IK calculation, selected manipulators for the test, the algorithms that are used to comparison. For example, El-Sherbiny et al. [11] used the Adaptive Neuro Fuzzy Inference System (ANFIS) algorithm to calculate the IK problem of a 5 DOF robot, and then compared results with GA algorithm. Both algorithms could get the appropriate solutions, but ANFIS algorithm proved to be the best one. The comparison also shows that a number of studies [12, 22, 23], using optimal algorithms such as PSO, ABC, Q-PSO ... handle the inverse kinetic requirements for the model of 7 degrees of freedom. All the used algorithms have proven the ability to handle the problem, but it is not difficult to see that most of these studies have the lower accuracy and processing speed than the ISADE as well as the Pro-ISADE algorithm proposed in this study.

Research	Robot arm	Results of Used algorithm	Results of Compared algorithm	Average of
Rokbani et al. [20]	3-DOF	10 Firefly	60 Firefly	
		1.27e−17	1.78e−18	Position error (m)
		1.21e−03	7.15e−3	Execution time (s)
Ayyıldız and Çetinkaya [7]	4-DOF	PSO	GA	
		7.70e−06	3.96e−04	Position error (m)
		0.0196	0.1753	Execution time (s)
El-Sherbiny et al. [11]	5-DOF	Adaptive Neuro Fuzzy Inference System (ANFIS)	GA	
		5.426e−03	7.64e−04	Position error (m)
		0.0308	83.1239	Execution time (s)
Shi and Xie [21]	6-DOF	Adaboost NN	—	
		0.00267	—	
		0.3	—	
Dereli and Köker [22]	7-DOF	Random IW-PSO	Global–Local Best IW-PSO	
		6.20e−03	3.64e−03	Position error (m)
		1.6	1.2	Execution time (s)
Serkan Dereli [12]	7-DOF	Q-PSO	PSO; ABC; Firefly	
		6.69347e-11	1.4547e-3	Position error (m)
		0.2195	0.4806	Execution time (s)
Serkan Dereli [23]	7-DOF	firefly	PSO, ABC	
		6.53e−05	5.45e−04	Position error (m)
		0,9204	0,4441	Execution time (s)
Our study	7-DOF	ISADE, Pro-ISADE	PSO, DE, Pro-PSO, Pro-DE	
		2.0103e-14	*7.4140e-13*	Position error (m)
		0.0554	*0.3604*	Execution time (s)

Italics were used to differentiate the results of Scenario 3.1 and 3.2.

Table 6.
Comparison with some other studies.

6. Conclusions

In this research, inverse kinematics problem for a 7 degree of freedom serial robot manipulator was implemented to prove the accuracy and efficiency of the self-adaptive control parameters in Differential Evolution (ISADE) and the ISADE algorithm with searching space improvement (Pro-ISADE) algorithm. To evaluate the effectiveness of the two algorithms above, the results obtained from the ISADE algorithm as well as Pro-ISADE were compared with the results from the PSO (Pro-PSO) and DE (Pro-DE) algorithm. Experiments were performed with three Scenarios. In the first Scenario, an endpoint in the workspace is randomly selected. The purpose of this Scenario is to compare the convergence speed of the three algorithms. In the second Scenario, algorithm was used to calculate inverse kinematics of the robot for 100 points randomly selected in the working space. The aim of this Scenario 2 is to test the accuracy and efficiency of the algorithm when the end effector started at the same position, it went to any point in working space. Meanwhile, in the third Scenario, the algorithms solved the inverse kinematics problem when the end effector of the robot moved point to point that are located on a spiral trajectory in the workspace. The implementation experiments have shown, the ISADE algorithm gave much better results than other algorithms in term of: accuracy, execution time and number of generation. Besides, by improving the searching boundary for joints' variable, the Pro-ISADE, Pro-DE and Pro-PSO also improve the accuracy as well as processing speed and especially the quality of the value of the joints variable compared to the ISADE, DE and PSO, respectively. These optimal joints' values ensure the feasibility of the dynamic and control problem in the future. In short, with ISADE algorithm as well as Pro-ISADE, they have handled the inverse kinematic requirement very effectively both in term of accuracy and computation time. The Pro-ISADE algorithm not only improves the above two factors, but also improves the quality of the joints' variables.

Conflict of interest

"The authors declare no conflict of interest."

Author details

Trung Nguyen[1*] and Tam Bui[1,2]

1 Hanoi University of Science and Technology, Hanoi, Vietnam

2 Shibaura Institute of Technology, Tokyo, Japan

*Address all correspondence to: trung.nguyenthanh@hust.edu.vn

IntechOpen

© 2021 The Author(s). Licensee IntechOpen. This chapter is distributed under the terms of the Creative Commons Attribution License (http://creativecommons.org/licenses/by/3.0), which permits unrestricted use, distribution, and reproduction in any medium, provided the original work is properly cited. CC BY

References

[1] Köker R, Çakar T. A neuro-genetic simulated annealing approach to the inverse kinematics solution of robots: a simulation based study. Eng Comput. 2016;32:553–565.

[2] Huang HC, Chen CP, Wang PR (2012) Particle swarm optimization for solving the inverse kinematics of 7-DOF robotic manipulators. IEEE international conference on systems, man, and cybernetics. Seoul, Korea, pp 3105–3110

[3] Rubio JJ, Bravo AG, Pacheco J, Aguilar C (2014) Passivity analysis and modeling of robotic arms. IEEE Lat Am Trans 12(8):1381–1389

[4] Kou YL, Lin TP, Wu CY (2014) Experimental and numerical study on the semi-closed loop control of a planar robot manipulator. Math Probl Eng 2014:1–9. doi:10.1155/2014/769038

[5] Carlos Lopez-Franco, Jesus Hernandez-Barragan, Alma Y. Alanis, Nancy Arana-Daniel. A soft computing approach for inverse kinematics of robot manipulators. Engineering Applications of Artificial Intelligence, 74, 104- 120, 2018.

[6] Mahanta GB, Deepak BBVL, Dileep M, et al. Prediction of inverse kinematics for a 6-DOF Industrial robot Arm using soft computing techniques. In: Soft computing for problem solving. Singapore: Springer; 2019. p. 519–530.

[7] Ayyıldız M, Çetinkaya K. Comparison of four different heuristic optimization algorithms for the inverse kinematics solution of a real 4-DOF serial robotmanipulator.Neural Comput Appl. 2016;27:825–836.

[8] Thomas Joseph Collinsm, Wei-Min Shen. Particle Swarm Optimization for High-DOF Inverse Kinematics. 3rd International Conference on Control, Automation and Robotics: IEEE, 2017.

[9] Malek Alkayyali, Tarek A. Tutunji, PSO-based Algorithm for Inverse Kinematics Solution of Robotic Arm Manipulators, 20th international Conference on Research and education in Mechatronics (REM), 2019

[10] Laura Cecilia Antonio-Gopar,Carlos Lopez-Franco∗, Nancy Arana-Daniel, Eric, Gonzalez-Vallejo and Alma Y. Alanis, Inverse Kinematics for a Manipulator Robot based on Differential Evolution Algorithm, IEEE Latin American Conference on Computational Intelligence (LACCI), 2018

[11] El-Sherbiny A, El-Hosseini MA, Haikal AY. A new ABC variant for solving inverse kinematics problem in 5 DOF robot arm. Appl Soft Comput. 2018;73:24–38.

[12] Serkan Dereli; Ra¸sit Köker; Ameta-heuristic proposal for inverse kinematics solution of 7-DOF serial roboticmanipulator: quantum behaved particle swarm algorithm; Artificial Intelligence Review, 2019

[13] Bui, T. Pham, H. Hasegawa, H., Improve self-adaptive control parameters in differential evolution for solving constrained engineering optimization problems. J Comput Sci Technol Vol.7, No.1 (2013), pp.59-74. DOI:10.1299/jcst.7.59.

[14] Coello, C.A.C., Use of a self-adaptive penalty approach for engineering optimization problems, Computers in Industry 41 (2000), pp.113-127.

[15] Thanh-Trung Nguyen, Ngoc-Linh Tao, Van-Tinh Nguyen, Ngoc-Tam Bui, V. H Nguyen, Dai Watanabe, Apply PSO Algorithm with Searching Space Improvements on a 5 Degrees of Freedom Robot, The 3rd International Conference on Intelligent Robotics and Control Engineering (IRCE 2020)

[16] Serkan Dereli, Ra¸sit Köker, Ameta-heuristic proposal for inverse kinematics solution of 7-DOF serial roboticmanipulator: quantum behaved particle swarm algorithm; Artificial Intelligence Review, 2019

[17] Craig JJ (2005) Introduction to robotics mechanics and control, 3rd edn. Pearson Education, Upper Saddle River, NJ

[18] Kennedy, J.; Eberhart, R. Particle swarm optimization. In: IEEE international conference on neural networks, (1995) pp 1943–1948

[19] Eberhart, R.; Kennedy, J. A new optimizer using particle swarm theory. In: The sixth international symposium on micro machine and human science, (1999) pp 39–4

[20] Rokbani N, Casals A, Alimi AM (2015) IK-FA, a new heuristic inverse kinematics solver using firefly algorithm. Comput Intell Appl Model Control 575:553–565

[21] ShiQ,Xie J (2017)Aresearch on inverse kinematics solution of 6-dof robot with offset-wrist based on adaboost neural network. In: IEEE international conference on CIS and RAM, 18–20 November, Ningbo, China

[22] Dereli S, Köker R (2018) IW-PSO approach to the inverse kinematics problem solution of a 7-DOF serial robot manipulator. Sigma J Eng Nat Sci 36:77–85

[23] Serkan Dereli & Raşit Köker, Calculation of the inverse kinematics solution of the 7-DOF redundant robot manipulator by the firefly algorithm and statistical analysis of theresults in terms of speed and accuracy, Inverse Problems in Science and Engineering, 2020, VOL. 28, NO. 5,

Chapter 3

Applying Improve Differential Evolution Algorithm for Solving Gait Generation Problem of Humanoid Robots

Van-Tinh Nguyen and Ngoc-Tam Bui

Abstract

This chapter addresses an approach to generate 3D gait for humanoid robots. The proposed method considers gait generation matter as optimization problem with constraints. Firstly, trigonometric function is used to produce trial gait data for conducting simulation. By collecting the result, we build an approximation model to predict final status of the robot in locomotion, and construct optimization problem with constraints. In next step, we apply an improve differential evolution algorithm with Gauss distribution for solving optimization problem and achieve better gait data for the robot. This approach is validated using Kondo robot in a simulated dynamic environment. The 3D gait of the robot is compared to human in walk.

Keywords: Humanoid robot, control data, differential algorithm, gait, optimization

1. Introduction

Humanoid robot is a complex machine with a series of joint and links. Biped motion asserts surpass advantage than the others due to flexibility and good adaption when moving on unpredictable surface. Difference from human beings, the humanoid robots have a limitation of structure and number of degree of freedom. Moreover, legged walking behavior requires an action of many active joints and is much more challenge in synthetic gait to keep balance. To solve this matter, the traditional approaches consider that ensuring Zero Moment Point within support polygon is important key. For example, Firstly, states of art is the works of Kajita [1–3] with the Linear Inverted Pendulum Model. Recently, Monje et al. [4] integrated the dynamic steadiness while moving in real time. Samadi and Moghadam-Fard [5] applied Gravity Compensated Inverted Pendulum Mode (GCIPM) with proposing that trajectory of ZMP is created by a first-order function. Kai Hu et al. [6] designed Compensative Zero-Moment Point Trajectory from the reference ZMP to decrease the effect of disturbances. Likewise, Yang et al. [7] presented bivariate-stability-margin-based control model to compensate zero moment point and modeling error, opposition-based learning algorithm is applied to generated gait pattern.

In the other direction, a number of researchers concentrate on central pattern generator (CPG)-based walking. Alongside the very common studies in [8–10], we can mention some up-to-date prominent instances such that JimmyOr [11] proposed an approach based on combination of CPG and ZMP to control spine motion, it is promising way to enable natural walk of the robot. Wei Li et al. [12] developed a mechanism to generate muscle forces for biped motion, this method designed a feedback controller which is formed by a dynamic neural network with CPGs.

A optimization-based approach considering stability and walking speed was introduced by Goswami Dip et al. [13]. Likewise, In-sik Lim et al. [14] addressed a gait generation technique for legged walking up and down stairs, in which, genetic algorithm and the human motion data were used to produce the optimal trajectory. Newly, Ho Pham Huy Anh and Tran Thien Huan [15] optimized walking gait by modified Jaya optimization algorithm.

Other scholars are interested in model-predictive methods. For example, Zamparelli [16] et al. designed a stable model predictive control to generate CoM trajectory for the robot on uneven surface. Scianca [17] presented a prediction model with stability constraints. Hildebrandt et al. [18] proposed a model-predictive approach for generating walking gait of the robots with redundant kinematic design.

In this chapter, we proposed a novel gait generation method, in which, trigonometric function with randomized coefficients is used to produce trial gait data for conducting simulation. The result is collected to build approximation function of output: rotation angle value, lateral and walking distance. Then, we designed an optimization problem with constraints and applied an improve differential algorithm for solving it. Optimal coefficients is returned to trigonometric function and define it exactly. By setting up cycle time, walking gait data will be generated by explicit trigonometric function and can be used for reference data in control.

Next part of this chapter is organized into five divisions: The first section introduces the subject of this research named Kondo KHR-3. The second section assigns gait functions to actuator joints. Thirdly, optimization problem with constraint is built to optimize gait trajectory. The novel differential evolution algorithm with Gauss distribution is developed in Section 4 to solve optimization problem. Section 5 includes results and discussions. The final section summarizes achievements of this research.

2. Biped model of Kondo

2.1 Physical configuration

The subject is presented in this chapter which is based on a humanoid robot named Kondo KHR-3 as depicted in **Figure 1a**. This robot consists of 22 degrees of freedom (DoF) However, we concentrated on lower body with 10 DoFs, in which 5 is the number of DoFs for each leg. The linkage configuration is shown in **Figure 1b**, where θ, d, m are rotation angle, length and weight of each link, respectively. Specially, the weigh of upper body is represented by m_o. The structure parameters of the robot are described in **Table 1**.

2.2 Foot structure with toe mechanism

Review on the works of foot structure for humanoid robots, we can list very old papers such as [19–21]. Recently, Sadedel et al. proposed a passive toe design. This mechanism improves energy efficiency of ankle and knee joints [22]. Likewise, Nerakae and Hasegawa simulated human-like foot structure which consist of big and tip toe to enhance biped walking gait in toe-off period [23]. In the other direction, Magistris et al. studied soft sole to minimize effect of the ground reaction force on stability during heel-strike period [24]. In [25], Daichi developed topology-

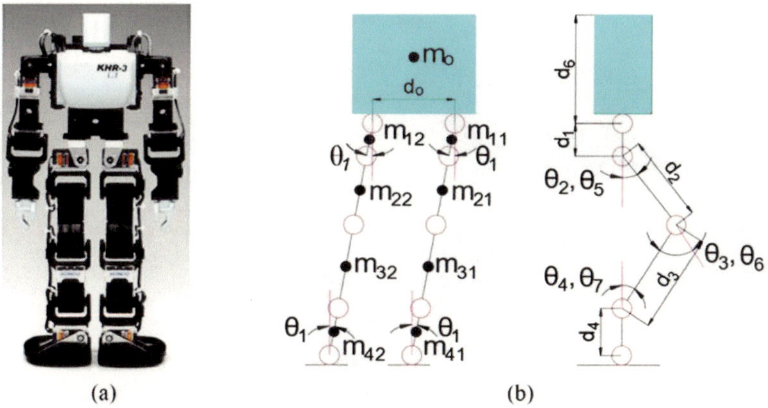

Figure 1.
Biped of Kondo: a) real robot; b) linkage model.

Parameters	Value
d_o	90.6 (mm)
d_1	38.5 (mm)
d_2	66.5 (mm)
d_3	62 (mm)
d_4	66 (mm)
d_5	168 (mm)
m_o	999.4 (g)
$m_{11,12}$	14 (g)
$m_{21,22}$	100 (g)
$m_{31,32}$	65 (g)
$m_{41,42}$	71.3 (g)

Table 1.
Structure parameters of humanoid robot.

based foot design to reduce weight and energy consumption, this research is specially meaningful for biped robot with limited physical parameters [26]. Our research presents a foot structure as shown in **Figure 2**, which is a combination of proposals mentioned in two papers [23, 26]. This sketch is built based on the following considerations: Firstly, all external forces acting on foot is at three areas consisting of heel, tip and big toe. Secondly, a passive toe joint mechanism with spring is applied to reduce impact of ground reaction force on foot. The material of foot is ABS plastic and design parameters are described in **Table 2a**.

2.3 Arm swing mechanism

Review on previous papers, we witness an arm swing mechanism which has been introduced and modeled using Adams in [27] as shown in **Figure 3**, the shoulder joint is linked to the hip joint by two linear springs. This mechanism is expected to generate the reaction moment from the arms to the trunk which eliminates ground reaction torque [28]. It makes the robot stable in motion. Our research applies this structure for the robot and characteristic parameters of the linear spring are stiffness of 0.8 $(newton/mm)$ and damping of 0.008 $(newton.sec\ /mm)$.

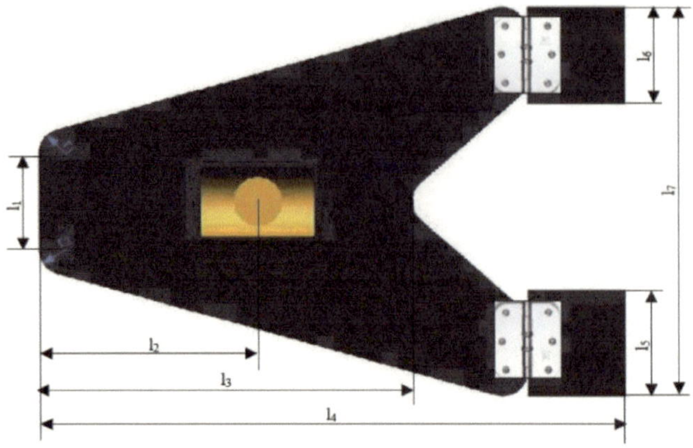

Parameters	Value
l_1	20 mm
l_2	40 mm
l_3	76.5 mm
l_4	120 mm
l_5	22 mm
l_6	22 mm
l_7	78 mm
r_1	5 mm
r_2	5 mm
Stiffness	2 (newton.mm/rad)
Damping	20 (newton-sec/mm)

Table 2a.
Design parameters

Figure 2.
Novel foot design.

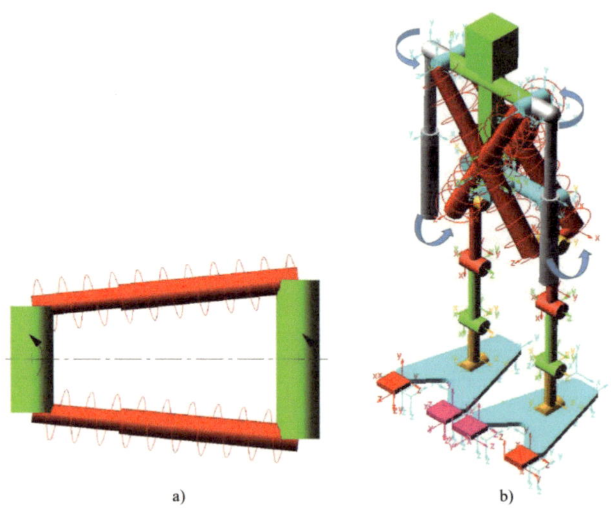

Figure 3.
Principle of arm swing: a) mechanism with linear spring; b) integration in simulation model.

3. Assignment of motion

The biped walking is a repeated physical action, thus, joint angle of the robot will follow up some circular principle. In this chapter, we adopt the trigonometric functions, which are the sine and the cosine, is widely applied for studying periodic problem. The general trigonometric function is proposed by Eq. (1). In motion, each leg of the robot performs walking behavior similarly to the other one, thus, we will use same gait function for each corresponding joint. In addition, the trajectory in sagittal plane of left leg is slower than right leg by $0.6s$ since a cycle is set to $1.2s$ while the hip and ankle joint gait data in frontal plane is identical for both of leg. We assign trajectories to all joints of the robot as described in Eq. (2)–(8). As can be seen that posture of the robot is defined at $0.3s$ in the beginning and stops motion at $3.3s$. $(0 - 0.3s)$ period is used for preparation and $(3.3 - 4.8s)$ period is used for checking stability.

$$\theta_i(t) = a_i + b_i \cos(\omega t) + c_i \sin(\omega t) + d_i \cos(2\omega t) \tag{1}$$

Where θ_i is the angle assigned to joint i, a_i, b_i, c_i, d_i are coefficients; t, i are time and index of joint, respectively; ω is an angular velocity, ω = 5.236 (rad/s)

$$\theta_1 = \begin{cases} 0; & t = 0 \; or \; t \geq 3.6 \\ \pm 1.5; & t = 0.3 \& t = 3.3 \\ \theta_1(t); & 0.3 < t < 3.3 \end{cases} \quad (2)$$

$$\theta_2 = \begin{cases} 0; & t \leq 0.3 \; or \; t \geq 3.6 \\ \theta_2(t + 0.6); & 0.3 < t < 3.3 \\ 15; & t = 3.3 \end{cases} \quad (3)$$

$$\theta_3 = \begin{cases} 0; & t \leq 0.3 \; or \; t \geq 3.6 \\ \theta_3(t + 0.6); & 0.3 < t < 3.3 \\ 30; & t = 3.3 \end{cases} \quad (4)$$

$$\theta_4 = \begin{cases} 0; & t \leq 0.3 \; or \; t \geq 3.6 \\ \theta_4(t + 0.6); & 0.3 < t < 3.3 \\ 15; & t = 3.3 \end{cases} \quad (5)$$

$$\theta_5 = \begin{cases} 0; & t = 0 \; or \; t \geq 3.3 \\ 15; & t = 0.3 \\ \theta_2(t); & 0.3 < t < 3.3 \end{cases} \quad (6)$$

$$\theta_6 = \begin{cases} 0; & t = 0 \; or \; t \geq 3.3 \\ 30; & t = 0.3 \\ \theta_3(t); & 0.3 < t < 3.3 \end{cases} \quad (7)$$

$$\theta_7 = \begin{cases} 0; & t = 0 \; or \; t \geq 3.3 \\ 15; & t = 0.3 \\ \theta_4(t); & 0.3 < t < 3.3 \end{cases} \quad (8)$$

4. Optimization problem with predictive function

4.1 Response surface methodology

Response surface model (RSM) is a mathematical model which is used for an approximation of stochastic process and is firstly introduced by Box and Wilson [27]. Our research adopts 3rd-order RSM to predict output value of simulation such as lateral and walking distance, rotation angle. The expression of 3rd-order RSM is displayed by Eq. (9).

$$\tilde{P}(x) = a_{po} + \sum_{i=1}^{n} b_{pi} x_i + \sum_{i=1}^{n} c_{pii} x_{ii}^2 + \sum_{ij(i<j)}^{n} c_{pij} x_i x_j + \sum_{i=1}^{n} d_{pii} x_{ii}^3. \quad (9)$$

Where n is a design variable number, n = 16, x_i is a design variable, and a_p, b_p, c_p, d_p are the coefficients of terms. Number of sampling for initialization is calculated by Eq. 10.

$$N_s = \frac{(n+1)(n+2)}{2} + n. \tag{10}$$

With $n = 16$, we determine $N_s = 169$ samples.

4.2 Optimization of gait trajectory

The goal of biped robot is to gain maximum walking distance with stable gait. This first problem is solved by a set of objective function, to apply optimization algorithm, we convert a maximum problem to minimum one as Eq. (11). The second issue is controlled by constraint functions as Eqs. (12) and (13).

$$Minimize\ f(x) = -Z_f \tag{11}$$

Subject to:

$$g_1(x) = 20 - |X_f| \geq 0 \tag{12}$$

$$g_2(x) = 5 - |R_f| \geq 0 \tag{13}$$

Where Z_f, X_f, R_f is walking, lateral distance and rotation angle, respectively. Their values are approximated by mentioned 3rd-order RSM.

5. Self-adaptive differential evolution with gauss distribution

This section introduce an modified self-adaptive differential evolution named G-SADE which is an improve version of Tam bui's algorithm (ISADE) proposed in [28].

5.1 Improve self-adaptive differential evolution

Reviewing on ISADE, it adopts an adaptive scaling factor of mutation process and adaptive mechanism for crossover control parameter. Scaling factor is generated by a sigmoid function with ranking individual in population. ISADE mainly consists of three operations.

5.1.1 Mutation operation

In [28], Tam Bui et al. randomly selected three mutation operation, which are DE/best/1, DE/best/2, and DE/rand to best/1. Among DE's operations, DE/best/1 and DE/best/2 are perceived for good convergence possession and DE/rand to best/1 is realized for good diversity property. Eqs. (14)–(16) shows the formula of chosen schemes.

$$DE/best/1: V_{i,j}^{iter} = X_{best,j}^{iter} + F*\left(X_{r1,j}^{iter} - X_{r2,j}^{iter}\right) \tag{14}$$

$$DE/best/2: V_{i,j}^{iter} = X_{best,j}^{iter} + F*\left(X_{r1,j}^{iter} - X_{r2,j}^{iter}\right) + F*\left(X_{r3,j}^{iter} - X_{r4,j}^{iter}\right) \tag{15}$$

$$DE/best/2: V_{i,j}^{iter} = X_{r1,j}^{iter} + F*\left(X_{best,j}^{iter} - X_{r1,j}^{iter}\right) + F*\left(X_{r2,j}^{iter} - X_{r3,j}^{iter}\right) \tag{16}$$

Where V, mutation vector; X, current vector; X_{best}, best vector in current population; $iter$, number of iterations; i, index of individual in population, NP; j, index of dimensions, D; $r1, r2, r3$, and $r4$ are different particles selected randomly from $[1; NP]$; and NP, D, number of individuals and problem dimensions, respectively. Scaling factor F is set to be high in the beginning iterations and after certain iteration, it automatically becomes smaller for proper exploitation. The value F is generated by a Sigmoid function as in Eq. (17).

$$F_i = \frac{1}{1 + exp\left(\alpha * \frac{i - \frac{NP}{2}}{NP}\right)} \quad (17)$$

Where α is the parameter that controls the value of scaling factor Fi as shown in **Figure 4**.

ISADE gives an addition scaling F_i^{mean} as in Eq. (18).

$$F_{iter}^{mean} = F_{min} + (F_{max} - Fmin)\left(\frac{iter_{max} - iter}{iter_{max}}\right)^{n_{iter}} \quad (18)$$

Where F_{max} and F_{min} are the lower and upper limitation of F, respectively. Recommended values for F_{max}, and F_{min} are 0.8 and 0.15, respectively. $iter_{max}$, and niter denote the maximum iteration, and the nonlinear modulation index. Niter is defined in Eq. (19).

$$n_{iter} = n_{min} + (n_{max} - n_{min})\left(\frac{iter}{iter_{max}}\right) \quad (19)$$

Where n_{max} and n_{min} are selected in the [0, 15] range. n_{max} and n_{min} are 0.2 and 6.0, respectively. Finally, scaling factor F_iter^i is calculated for each particle in each iteration as Eq. (20). The particle in population will receive different scaling factor. This mechanism maintains the balance of exploration and exploitation process in mutation operation.

$$F_{iter}^i = \frac{F_i + F_{iter}^{mean}}{2} \quad (20)$$

5.1.2 Crossover operation

After mutation operation, ISADE adopts a crossover operation as Eq. (21).

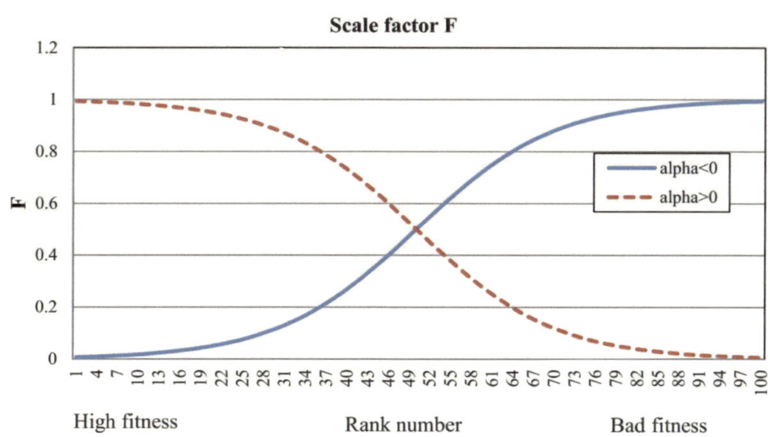

Figure 4.
Value of scale factor F depends on rank number i and α.

$$U_{i,j}^{iter} = \begin{cases} V_{i,j}^{iter} & \text{if } rand[0,1] \leq CR \; or \; j = j_{rand} \\ X_{i,j}^{iter} & \text{otherwise.} \end{cases} \tag{21}$$

Where U, V, and X are a trial vector, mutation vector, and current vector, respectively. CR is a crossover parameter within the interval (0,1). j_{rand} is an integer random number selected from set {1, 2, ..., D}.

5.1.3 Selection operation

This stage will choose the better fitness between trial vector U and target vector X for the next generation as Eq. (22).

$$X_i^{iter+1} = \begin{cases} U_i^{iter} & \text{if } f\left(U_i^{iter}\right) \leq f\left(X_i^{iter}\right) \\ X_i^{iter} & \text{otherwise.} \end{cases} \tag{22}$$

5.2 Self-adaptive differential evolution with gauss distribution

Gauss distributions play an important role in statistics and are often used to generate real-valued random variables. Gaussian distribution gives a fantastic possibility to control the exploiting zone in optimization based on complexity of each considered problem. Our proposed strategy generates a new trial vectors in mutation operation by multiply Gauss random variable to a vector difference $(X_{r1,j}^{iter} - X_{r2,j}^{iter})$ in Eq. (14) and $(X_{r3,j}^{iter} - X_{r4,j}^{iter})$ in Eq. (15).

The formula of schemes in mutation operation is modified as the following Eq. (23) and (24).

$$DE/best/1: V_{i,j}^{iter} = X_{best,j}^{iter} + N(0,\sigma^2) * F * \left(X_{r1,j}^{iter} - X_{r2,j}^{iter}\right) \tag{23}$$

$$DE/best/2: V_{i,j}^{iter} = X_{best,j}^{iter} + N(0,\sigma^2) * \left[F * \left(X_{r1,j}^{iter} - X_{r2,j}^{iter}\right) + F * \left(X_{r3,j}^{iter} - X_{r4,j}^{iter}\right)\right] \tag{24}$$

Where $N(0,\sigma^2)$ is a Gauss random variable with mean *zero* and standard deviation σ. Recommended value for σ is 0.1. The new proposed DE is implemented as in Pseudocode 1 and The optimization process is described in Pseudocode 2.

Pseudocode 1: Implement of proposed DE

1. **Initialization**: Generating randomly NP initial populations inside the bounds.
2. **Evaluation and ranking population**: Evaluating and ordering all individuals based on their fitness value. Next, the first best NP individuals are selected to survive and kept for the next generation.
3. **Adaptive scaling factor**: Computing scale factors F by Eqs. (17) to (20).
4. **Mutation operation**: Mutation vectors V are created by applying adaptive selection learning strategy through Eq. (23) to (24).
5. **Crossover operation**: Calculating trial vector U using Eq. (21).
6. **Selection Operation**: Comparing trial vector U_i^{iter} to target vector X_i^{iter} is implemented based on evaluating their fitness value.
   The better candidate will be chosen for propagating new offspring in the next generation by Eq. (22).
7. Steps 2 to 6 is repeated while termination condition is checked

Pseudocode 2: Implement of optimization

1. Achieving the first simulation in Adams by Trial and Error method.
2. Design of experiment.
3. While termination condition is not reached;
   Making 3rd order response surface model;
   Applying ISADE with objective function 11 and constraint function 12, 13
   Analyzing the result through dynamic simulation in Adams.
4. The optimal value of design variable is achieved

6. Results and discussions

Surfaces shown in **Figure 5** is employed to present the walking process of biped robot. This is a absolutely flat terrain.

To observe the effect of arm swing mechanism on walking behavior, we build the 3D robot model in dynamic simulated environment of Adams software. Two configurations of the robot with different arm mechanism are considered as the following: firstly, we lock shoulder joint of the arm or in other respects the upper body is a static block with no movement. Second configuration is designed with a 1-32DoF shoulder joint which is set in sagittal plane. The biped motion is studied and evaluated on an above-mentioned environment. After applying G-SADE algorithm to solve optimization problem, we gain the optimal design variable is presented in **Table 2.** By replacing the term coefficients of i, a_i, b_i, c_i, d_i in Eq. (1) with these optimal value, we will define four gait functions for all joints of the legs and trajectory is ruled by the principle shown by Eqs. (2)–(8). To be clear, we can see these gait patterns in **Figure 6**. After all, results of simulation are presented in **Figure 7**.

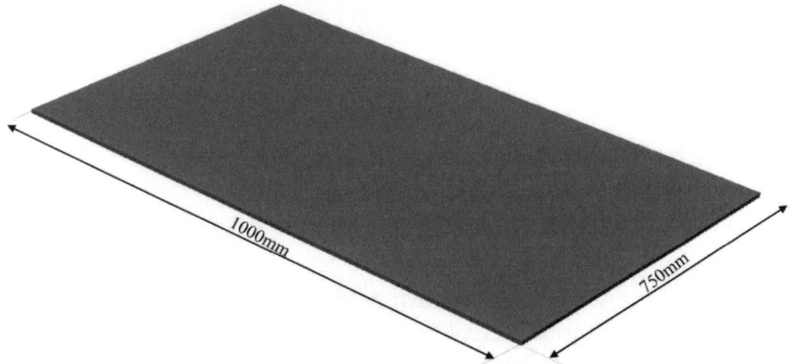

Figure 5.
Considered surface for simulation.

i	Optimal design variables.			
	a_i	b_i	c_i	d_i
1	0.0034	0.0446	0.0011	−0.0070
2	0.0410	0.2104	0.0007	0.0134
3	0.4138	0.0388	−0.1894	−0.1108
4	−0.2453	0.1460	0.1031	0.0493

Table 2.
Design variable value.

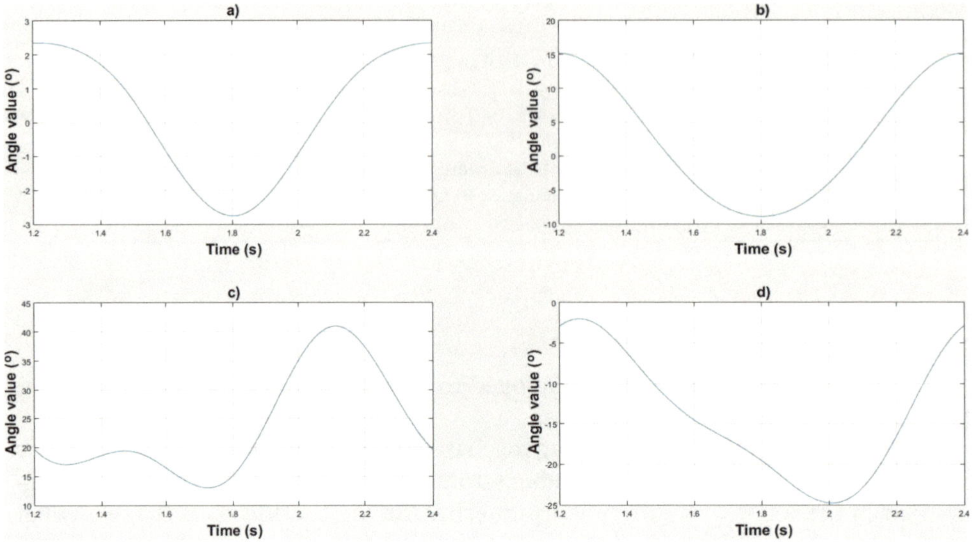

Figure 6.
Gait pattern: (a) hip joint angle in frontal plane; (b) hip joint angle in sagittal plane; (c) knee joint angle; (d) ankle joint angle.

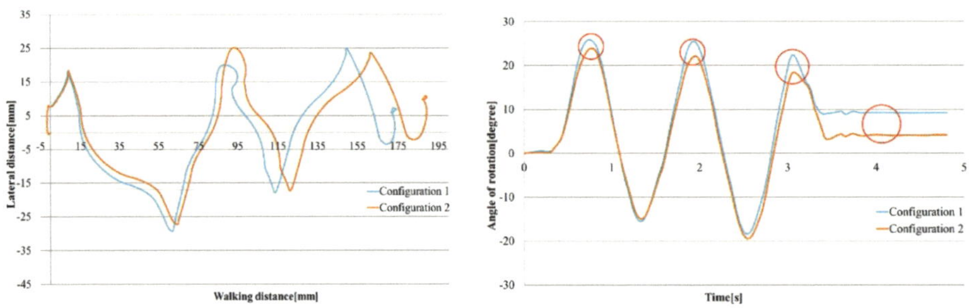

(a) (b)

Figure 7.
Simulation result: (a) CoM trajectory in horizontal plane; (b) rotation angle of robot.

After implementing the simulation with achieved control data, we attain the final position of the robot as the following: For configuration 1 with no shoulder joint, lateral distance of X_f, walking distance of Z_f and rotation angle of R_f are 6.17(mm), 172.11(mm), and 9.19^o, respectively; meanwhile these value are 9.21(mm), 184.66(mm), and 3.98^o, respectively for configuration 2 with 1-DoF shoulder joint.

To be particular, **Figure 7a** depicts the CoM trajectory of the robot, which has a approximately periodic waveform. Comparing to the humans described in [29], this performance is similar. Moreover, model with arm swing mechanism witnesses an improvement of about 9% in walking distance from 172.11(mm) to 184.66(mm).

Figure 7b presents the rotation angle of the robot depending on time axis of the motion process. As can be seen that this line chart undergoes a fluctuation about from -25° to 25° around center line. It means that the robot's feet rotate significantly in the locomotion. This phenomenon is due to an impact of friction force between foot and ground. Furthermore, the angle rotation of configuration 2 fluctuates with smaller amplitude of 5%, at the final position, this angle decreases by about 55%. (3.0s - 3.3s) period is prepared for landing and checking stability then, the rotation angle of the robot rapidly declines. In (3.3s - 4.8s) period, this angle is constant which expresses the success of the simulation and the robot stands steadily on the ground. Totally, above discussions address that the model with 1-DoF shoulder joint performs a better exhibition and should be applied in real robot.

Next discussion is to compare 3D gait on flat ground of both configurations with the humans in a cycle. The walking behavior is shown in **Figure 8**. As can be seen that posture of the simulation model at $1.2s, 1.5s, 1.8s, 2.1s$ is comparable to the human walking behavior in $"initial - contact"$, $"mid - stance"$, $"terminalstance"$, and $"toe - off"$ periods. Highlight points on the robot posture are $"toe - off"$ periods at $1.5s$ and $2.1s$. This behavior is very clear and has pros on the contact between foot and ground in phase transferring stage.

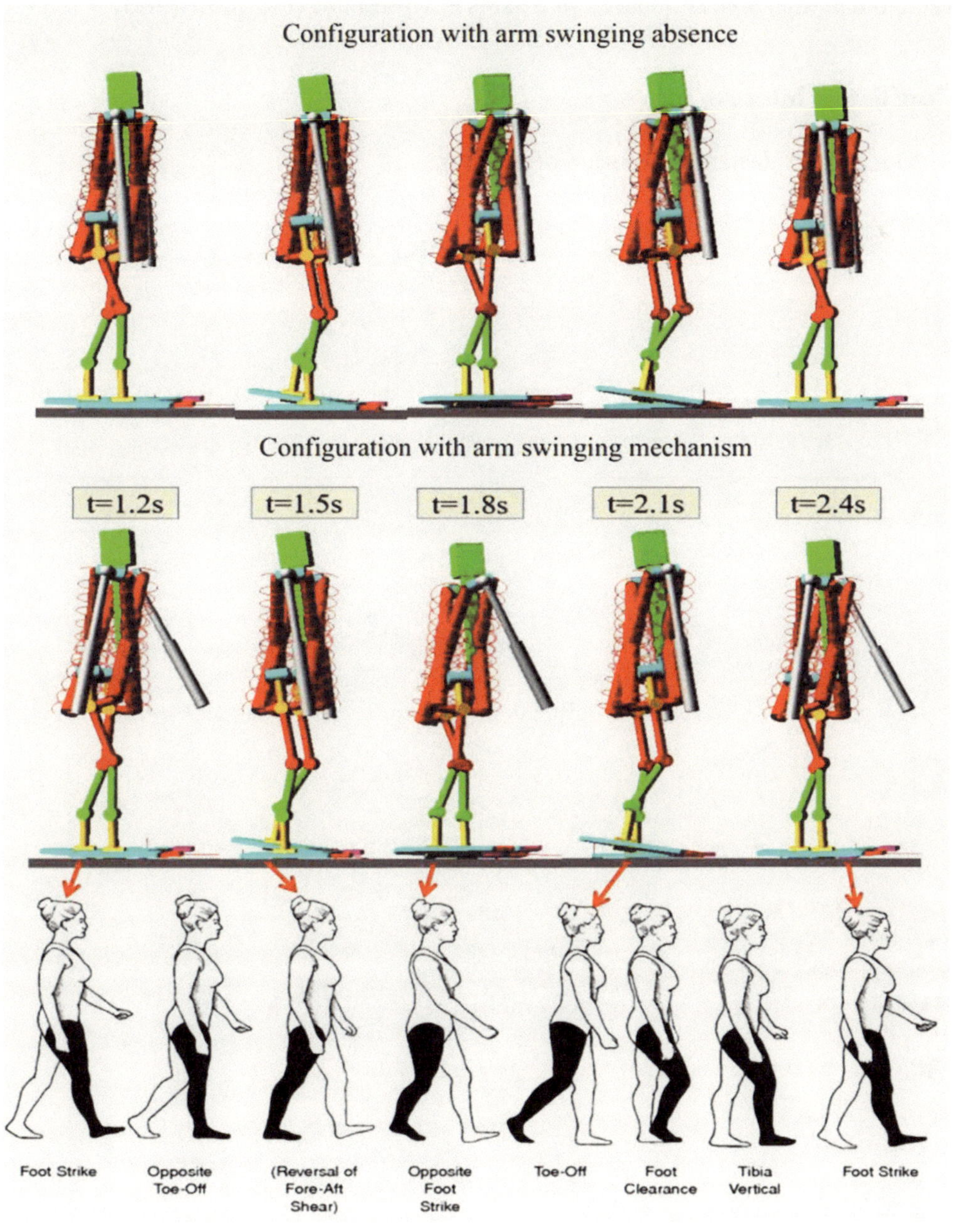

Figure 8.
3D gait in a cycle.

7. Conclusions

Trajectory generation matter for biped walking is always one of the most challenges. Thus, researches on this field is very attractive and is implemented ceaselessly until now. In the same direction, our chapter one again relates to an 3D gait generation method for the biped robot. This approach constructs an optimization problem with constraint to create gait data of the robot. In addition, the modified differential evolution algorithm is introduced to solve an optimization problem. In the next stage, this chapter confirms that the arm swing mechanism enhances the biped walking performance by improving walking distance and reducing rotation angle when the robot moves on the flat ground. The effectiveness of this method is validated by dynamic simulation in Adams environment (MSC software).

Conflict of interest

The authors declare no conflict of interest.

Author details

Van-Tinh Nguyen* and Ngoc-Tam Bui
Hanoi University of Science and Technology, Hanoi, Vietnam

*Address all correspondence to: tinh.nguyenvan@hust.edu.vn

IntechOpen

© 2021 The Author(s). Licensee IntechOpen. This chapter is distributed under the terms of the Creative Commons Attribution License (http://creativecommons.org/licenses/by/3.0), which permits unrestricted use, distribution, and reproduction in any medium, provided the original work is properly cited.

References

[1] Kajita, S., and Tanie, K. (1991). "Study of dynamic biped locomotion on rugged terrain – derivation and application of the linear inverted pendulum mode", in Robotics and Automation, 1991. Proceedings. 1991 IEEE International Conference on, Vol. 2 (Sacramento, CA: Institute of Electrical Engineers, IEEE), 1405–1411.

[2] Kajita, S., Kanehiro, F., Kaneko, K., Yokoi, K., and Hirukawa, H. (2001). "The 3d linear inverted pendulum mode: a simple modeling for a biped walking pattern generation," in Intelligent Robots and Systems, 2001. Proceedings. 2001 IEEE/RSJ International Conference on, Vol. 1 (Maui: Institute of Electrical Engineers, IEEE), 239–246

[3] Kajita, S., Kanehiro, F., Kaneko, K., Fujiwara, K., Harada, K., Yokoi, K., et al. (2003). "Biped walking pattern generation by using preview control of zero moment point", in Robotics and Automation, 2003. Proceedings. ICRA '03. IEEE International Conference on, Vol. 2 (Taipei: Institute of Electrical Engineers, IEEE), 1620–1626.

[4] Majied Mokhtari, Mostafa Taghizadeh, Mahmood Mazare (2021). "Impedance control based on optimal adaptive high order super twisting sliding mode for a 7-DOF lower limb exoskeleton", in Meccanica 56:3, pages 535-548.

[5] Samadi, Farshad; Moghadam-Fard, Hamid (2014). "Pattern generation for humanoid robot with natural ZMP trajectory", in IEEE 2014 Second RSI/ ISM International Conference on Robotics and Mechatronics (ICRoM) - , 570–575, Tehran, Iran

[6] K. Hu, C. Ott and D. Lee (2016). "Learning and Generalization of Compensative Zero-Moment Point Trajectory for Biped Walking", in IEEE Transactions on Robotics, vol. 32, no. 3, pp. 717-725, June 2016, doi: 10.1109/ TRO.2016.2553677.

[7] Liang Yang, Zhi Liu, Yong Chen (2018). "Bipedal walking pattern generation and control for humanoid robot with bivariate stability margin optimization", in Advances in Mechanical Engineering, Volume 10 issue: 9.

[8] Endo G, Morimoto J, Matsubara T, Nakanishi J, Cheng G. (2008). "Learning CPG-based Biped Locomotion with a Policy Gradient Method: Application to a Humanoid Robot", in the International Journal of Robotics Research;27(2):213-228. doi: 10.1177/0278364907084980

[9] Takeshi Mori, Yutaka Nakamura, Masa-aki Sato, Shin Ishii (2004). "Reinforcement Learning for a CPG-driven Biped Robot", in AAAI'04: Proceedings of the 19th national conference on Artifical intelligence, pp. 623–630.

[10] Yutaka Nakamura, Takeshi Mori, Masa-aki Sato, Shin Ishii (2007). "Reinforcement learning for a biped robot based on a CPG-actor-critic method", in Neural Networks, Volume 20, Issue 6, Pages 723-735

[11] Jimmy Or (2010). "A hybrid CPG–ZMP control system for stable walking of a simulated flexible spine humanoid robot", in Neural Networks, Volume 23, Issue 3, Pages 452-460.

[12] Li W., Szczecinski N.S., Hunt A.J., Quinn R.D. (2016). "A Neural Network with Central Pattern Generators Entrained by Sensory Feedback Controls Walking of a Bipedal Model", In: Lepora N., Mura A., Mangan M., Verschure P., Desmulliez M., Prescott T. (eds) Biomimetic and Biohybrid Systems. Living Machines 2016. Lecture

Notes in Computer Science, vol 9793. Springer, Cham. https://doi.org/10.1007/978-3-319-42417-0_14

[13] Dip, G., Prahlad, V., & Kien, P. (2009). "Genetic algorithm-based optimal bipedal walking gait synthesis considering tradeoff between stability margin and speed", in Robotica, 27(3), 355-365. doi:10.1017/S026357470800475X

[14] In-sik Lim, Ohung Kwon, Jong Hyeon Park (2014). "Gait optimization of biped robots based on human motion analysis", in Robotics and Autonomous Systems, Volume 62, Issue 2, Pages 229-240,

[15] Ho Pham Huy Anh and Tran Thien Huan (2020). "Optimal Walking Gait Generator for Biped Robot Using Modified Jaya Optimization Technique", in International Journal of Computational Intelligence Systems, Volume 13, Issue 1, pp. 382 - 399.

[16] Alessio Zamparelli, Nicola Scianca, Leonardo Lanari, Giuseppe Oriolo (2018). "Humanoid Gait Generation on Uneven Ground using Intrinsically Stable MPC", in IFAC-PapersOnLine, Volume 51, Issue 22, pp. 393-398.

[17] Hildebrandt, AC., Schwerd, S., Wittmann, R. et al. (2019). "Kinematic optimization for bipedal robots: a framework for real-time collision avoidance", in Auton Robot 43, pp. 1187–1205.

[18] R. Sellaouti, O. Stasse, S. Kajita, K. Yokoi, and A. Kheddar (2006). "Faster and smoother walking of Humanoid HRP-2 with passive toe joints", in Proceedings of IEEE/RSJ international conference on intelligent robots and systems, pp. 4909-4914.

[19] S. Lohmeier, T. Buschmann, H. Ulbrich, and F. Pfeiffer (2006). "Modular joint design for performance enhanced humanoid robot LOLA", in Proceedings of IEEE international conference on robotics and automation, pp. 88-93.

[20] K. Yamane and L. Trutoiu (2009). "Effect of Foot Shape on Locomotion of Active Biped Robots, a study on effect of two-arch structure of foot for biped robots", Proceedings of 9th IEEE-RAS International Conference on Humanoid Robots, pp.BIBLIOGRAPHY 74 2230-2236.

[21] Sadedel, M., Yousefi-Koma, A., Khadiv, M., & Mahdavian, M. (2017). "Adding low-cost passive toe joints to the feet structure of SURENA III humanoid robot", in Robotica, 35(11), 2099-2121. doi:10.1017/S026357471600059X

[22] K. Nerakae and H. Hasegawa (2014). "Big toe sizing design of small biped robot by using gait generation method",in Applied Mechanics and Materials, volume 541-542, pp. 1079-1086.

[23] Giovanni de Magistris, Sylvain Miossec, Adrien Escande, Abderrahmane Kheddar (2017). "Design of optimized soft soles for humanoid robots", in Robotics and Autonomous Systems, Elsevier, 95, pp.129-142.

[24] K. Daichi (2016). "Optimization of the biped robot for stable walking on rough terrain", in Master's thesis, Shibaura Institute of Technology.

[25] F. Naoki (2017). "Energy saving of biped robots aiming for 24-hour continuous operation", in Master's thesis, Shibaura Institute of Technology, Japan.

[26] J. Park (2008). "Synthesis of natural arm swing motion in human bipedal walking", in Journal of Biomechanics, vol. 41, No. 7, pp. 1417-1426.

[27] Box, G. E. P., & Wilson, K. B. (1951). "On the experimental

attainment of optimum conditions", in Journal of the Royal Statistical Society, Series B (Methodological), Vol. XIII, No. 1.

[28] T. Bui, H. Pham, and H. Hasegawa (2013). "Improve self-adaptive control parameters in differential evolution for solving constrained engineering optimization problems", Journal of Computational Science and Technology, vol. 7, no. 1, pp. 59-74.

[29] M. S. Orendurff, A. D. Segal, J. S. Berge, K. C. Flick, D. Spanier, and G. K. Klute (2006). "The kinematics and kinetics of turning: limb asymmetries associated with walking a circular path", Gait and Posture, vol. 23, pp. 106-111.

Chapter 4

QoS Control in Remote Robot Operation with Force Feedback

Pingguo Huang and Yutaka Ishibashi

Abstract

Recently, many researchers focus on studies of remote robot operation with force feedback. By using force feedback, since users can touch remote objects and feel the shape, weight, and softness of each object, the efficiency and accuracy of operation can be largely improved. However, when the haptic information such as force and/or position information is transmitted over a QoS (Quality of Service) non-guaranteed network like the Internet, QoE (Quality of Experience) and stability may seriously deteriorate. Therefore, it is important to carry out QoS control and stabilization control together to solve the problems. In this chapter, we mainly focus on QoS control. We also introduce our remote robot system with force feedback which we constructed to study QoS control and stabilization control by experiment. In the system, a user operates a remote industrial robot with a force sensor by using a local haptic interface device while monitoring the robot operation by a video camera. We handle two types of operation; operation with a single remote robot system and that between two remote robot systems. We explain several types of QoS control which we have proposed so far for remote robot operation with force feedback. Finally, we discuss the challenges and future directions of QoS control in remote robot operation with force feedback.

Keywords: remote robot operation, force feedback, haptic interface device, QoS control

1. Introduction

Recently, many researchers focus on studies of remote robot operation with force feedback in which a user operates a remote robot having force sensors by using a haptic interface device while monitoring the remote operations by a video camera [1–3]. By using force feedback, since users can touch remote objects and feel the shape, weight, and softness of each object, the efficiency and accuracy of operation can be largely improved [4]. Therefore, the remote robot operation with force feedback is expected to be used in many areas such as remote surgery, disaster rescue, and outer space. However, when the information about force and/or position is transmitted over a QoS (Quality of Service) [5] non-guaranteed network like the Internet, QoE (Quality of Experience) [6] may seriously deteriorate [3, 4] owing to the network delay, delay jitter, and packet loss. Furthermore, as the network delay increases, the reaction force becomes larger, and unstable phenomena such vibrations of the robot and device may occur more often [7–9]. To solve the problems, we need to carry out QoS control and stabilization control together [4]. In this chapter, we mainly focus on QoS control at the application layer. The QoS control alleviates the influences of network delay, delay jitter, and packet loss on QoE.

IntechOpen

We also introduce our remote robot system with force feedback which we constructed to study the QoS control and stabilization control by experiment. In the system, a user operates a remote industrial robot with a force sensor by using a local haptic interface device while monitoring the robot operation. We handle two types of operation; operation with a single remote robot system and that between two remote robot systems. We explain several types of QoS control which we have proposed so far for remote robot operation with force feedback.

In this chapter, first, we explain the remote robot system with force feedback in Section 2. Next, we introduce expected applications of the remote robot system with force feedback in Section 3. Then, we outline the problems to be solved for the applications in Section 4 and describe the QoS control which is used to solve the deterioration problems owing to the network delay, delay jitter, and packet loss in Section 5. Finally, we discuss the challenges and future directions of QoS control in Section 6 and conclude the chapter in Section 7.

2. Remote robot system with force feedback

2.1 System configuration

The configuration of the remote robot system with force feedback is shown in **Figure 1**. The system consists of two terminals called the master terminal and slave terminal. Each terminal consists of two PCs, and the PCs are connected to each other via a switching hub.

At the master terminal, a 3 DoF (Degree of Freedom) haptic interface device (3D Systems Touch [10]) is connected to PC for haptic interface device, and another PC is used for video. At the slave terminal, one of the two PCs is used for a web camera (produced by Microsoft Corp., and video resolution is 1920 × 1080 pixels), and the other PC is used for industrial robot. The industrial robot consists of a 6 DoF robot arm (RV-2F-D by Mitsubishi Electric Corp. [11]), a robot controller (CR750-Q [11]), and a force sensor (1F-FS001-W200 [12]). The force sensor is attached to the surface of the flange of the robot arm. The force sensor is connected to the robot controller via the force interface unit.

2.2 Remote operation

A user at the master terminal can operate the industrial robot at the slave terminal by using the haptic interface device while watching video (coding scheme:

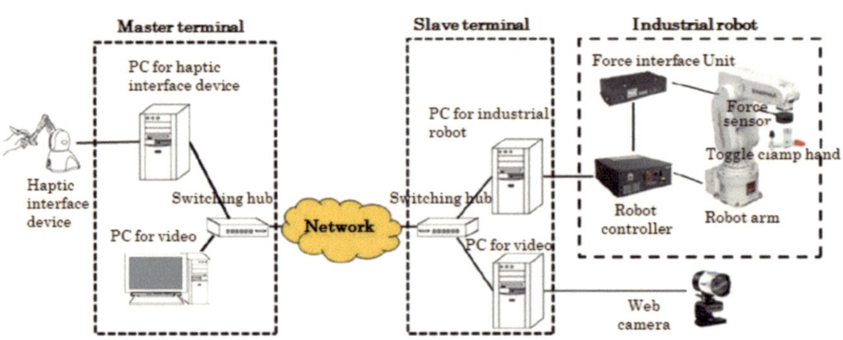

Figure 1.
Configuration of remote robot system with force feedback.

Motion JPEG, average bit rate: 4.5 Mbps). The default position of the haptic interface device is set to the origin, and the position corresponds to the default position of the industrial robot [13].

The master terminal updates the position information, calculates the reaction force, and outputs the reaction force every millisecond. The master terminal also transmits the position information to the slave terminal by User Datagram Protocol (UDP). At the slave terminal, the command information which is based on the position information received from the master terminal is sent to the industrial robot every 3.5 milliseconds by the real-time control function [14]. The force information is also acquired by the real-time control function, and the information is transmitted to the master terminal by UDP.

The reaction force $\boldsymbol{F}_t^{(\mathrm{m})}$ applied to the haptic interface device at time t ($t \geq 1$) is calculated as follows:

$$\boldsymbol{F}_t^{(\mathrm{m})} = K_{\mathrm{scale}} \boldsymbol{F}_{t-1}^{(\mathrm{s})} \tag{1}$$

where $\boldsymbol{F}_{t-1}^{(\mathrm{s})}$ denotes the force received from the slave terminal (note that we use only 3 DoF of force here), and K_{scale} is a force scale which is set to 1 in this paper. Furthermore, since the maximum force applied to the haptic interface device is 3.3 N [10], the reaction force is set to 3.3 N when the calculated force is larger than 3.3 N.

The position vector $\boldsymbol{S}_t$ of the industrial robot outputted at the time t ($t \geq 2$) is calculated as follows:

$$\boldsymbol{S}_t = \boldsymbol{M}_{t-1} + \boldsymbol{V}_{t-1} \tag{2}$$

where $\boldsymbol{M}_t$ is the position vector of haptic interface device received from the master terminal at time t, $\boldsymbol{V}_t \left(= \boldsymbol{M}_t - \boldsymbol{M}_{t-1}\right)$ is the velocity vector and $|\boldsymbol{V}_t| \leq V_{\max}$, and $V_{\max}$ is the maximum movement velocity. That is, in order to operate the robot arm safely, the maximum movement velocity is limited to $V_{\max}$ ($V_{\max}$ = 5 mm/s [13] in this chapter).

In this chapter, we handle two types of operation, operation with single remote robot system and that between two remote robot systems. In the latter operation, we deal with two types of work (carry together and hand delivery). In carry together, two industrial robots carry an object together. In hand delivery, an object was hand-delivered between the two industrial robots.

3. Expected applications

As shown in **Figure 2**, the remote robot system with force feedback is expected to be used in various areas.

3.1 Remote surgery/rehabilitation

In order to solve the problems of imbalance of medical resources, remote surgery/rehabilitation using the remote robot system with force feedback is an effective method. Also, the system can be used for remote surgery training for medical interns.

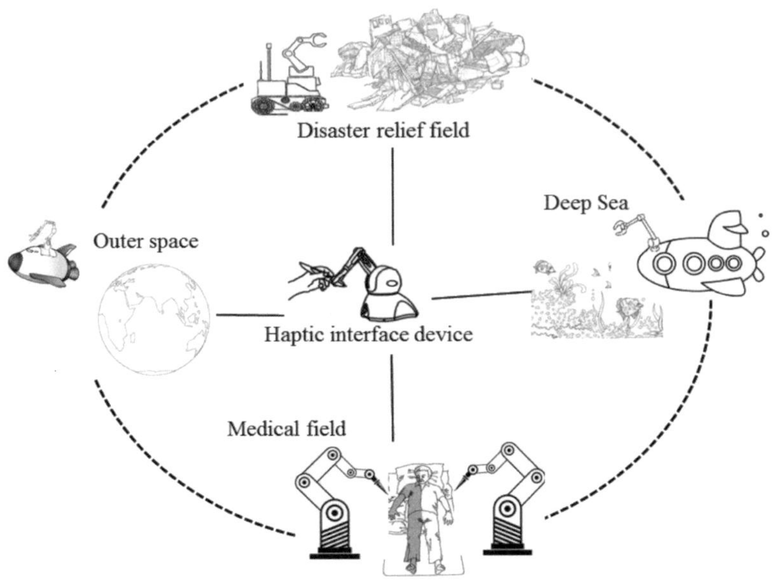

Figure 2.
Expected applications.

3.2 Work in dangerous areas

It is difficult for human to work in danger areas such as deep sea and outer space. Therefore, we can employ robots to work in the danger areas instead of humans. By using the remote robot system with force feedback, we can control the robot which works in danger areas from a remote safe area. We can improve the efficiency and accuracy of work by force feedback.

3.3 Disaster rescue and relief

The remote robot system with force feedback can also be used for rescue and relief from disasters such as earthquake and concentrated downpour. In this case, the remote robot can be rescue robot or drone, and the system can be used to help people, to distribute goods for disaster victims. Also, it can be used to confirm disaster situation.

In these applications, it is difficult for only robots to work because the situations are unknown in advance. Thus, human's support is needed. This means that we need robots to help humans and robots also need human's supports. Therefore, the cooperation among humans and robots is needed.

4. Problems to be solved

In this section, we explain the problems to be solved for widespread applications of the remote robot system with force feedback.

4.1 Problem of cooperation

There exist several types of cooperation using the remote robot system with force feedback, for example, cooperation between human and human, robot and human, robot and robot. In the cooperation between human and human, the will transmission between humans is important, and efficient transmission methods should be established for the cooperation. In the cooperation between human and robot, it is necessary to consider whether humans support robots or robots support humans, and how to support each other more efficiently. In the cooperation between robot and robot, we have two cases; in one case, the robots can cooperate by communications between the slave terminals; in the other case, they can do with force sensors because they are connected to each other through an object when they carry or hand-deliver the object. The cooperation is important when speedy control is needed.

4.2 Problem of will transition

In the cooperation between human and human, it is important to transmit users' will (for example, movement directions and speeds) to each other, and will transmission using haptic may reduce the transmission time, and it is possible to transmit wills in delicate manipulation work in which it is difficult to transmit wills only by traditional methods (i.e., wills transmitted by audio and video) [15]. Therefore, it is necessary to establish an efficient method to transmit/determine wills by haptic for the cooperation.

4.3 Network delay, delay jitter, and packet loss

As described in Section 1, when the information about force and/or position transmitted over a QoS non-guaranteed network like the Internet, the reaction force may become large and QoE may be seriously deteriorate owing to the network delay, delay jitter, and packet loss. It is necessary to carry out QoS control to solve the problems.

4.4 Unstable phenomena

In the remote robot system with force feedback, the system may be unstable since there exists a control loop between a haptic interface device and a remote robot (see **Figure 3**). As the network delay increases, the movement of the remote robot is largely later than that of the haptic interface device, the reaction force becomes larger, and unstable phenomena such vibrations of the robot and device may occur more often. Also, in remote cooperation, there exists more loops in the cooperation systems and the unstable phenomena become more complex and difficult problems [16]. It is important to carry out stabilization control to solve the problems.

4.5 Cooperation in case of emergency

In the remote cooperation, there exist emergency cases in which network interruption occurs and users may be not able to control remote robots to do

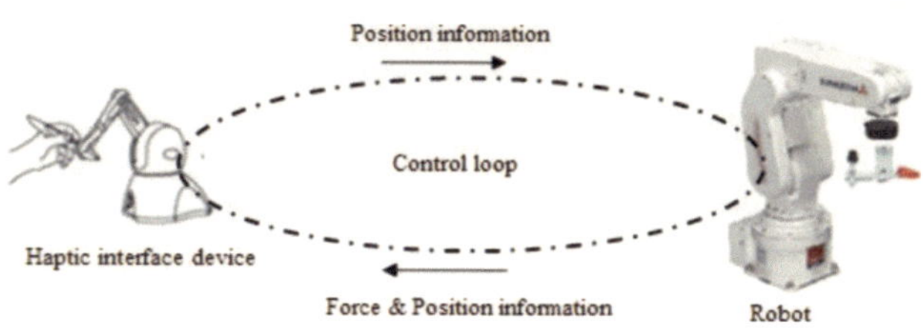

Figure 3.
Control loop between haptic interface device and robot.

collaborative work. In this case, remote robots need to intercommunicate with each other and finish cooperation based on force sensors. That is, we need to handle the case of emergency and establish effective methods in the case.

From the above, there are many problems to be solved. Here, we mainly focus on QoS control which alleviates the influences of network delay, delay jitter, and packet loss.

5. QoS control

QoS control is effective for solving the problems occurred by network delay, delay jitter, and packet loss. As described in the previous section, in the remote robot operation, there exists a control loop between a haptic interface device and a robot. We need to carry out QoS control in the loop to improve the QoE. This means that we need to carry out QoS control at a haptic interface device terminal and/or at a robot terminal. There are many types of QoS control such as traffic management and control, error control, spatiotemporal synchronization control (we can carry out media synchronization control or causality control to achieve spatiotemporal synchronization), consistency control, adaptive reaction force control [4], and position control using force information [17]. We also introduce serval types of QoS control which we previously proposed for the remote robot operation.

5.1 Media synchronization control

Media synchronization control is used to solve the problems occurred by network delays and delay jitter. The control can be grouped into intra-stream synchronization control, inter-stream synchronization control, and inter-destination (or group) synchronization control [18].

Intra-stream synchronization control is used to preserve the timing relation between media units (MUs, which are information units for media synchronization) [19] in a single media stream. There are several types of intra-stream synchronization control, for example, Skipping [19], Virtual-Time Rendering (VTR) [19], and so on. Skipping outputs MUs on receiving the MUs, and when the sequence number of a received MU is smaller than that of the last-output MU, the control discards the received MU. VTR has a virtual-time axis which can be contracted or expanded dynamically according to the network delay, and MUs are output along the virtual-time axis.

In multimedia applications, if we only carry out intra-stream synchronization control for each media stream separately, the temporal relationship among media streams may be disturbed and QoE may be deteriorated. In order to solve the problem, we need to carry out inter-stream synchronization control. The VTR can be used for intra-stream and inter-stream synchronization control. Under the control, one media stream is handled as the master stream and the others are dealt with as slave streams. VTR carries out only the intra-stream synchronization control for the master stream, and it exerts the inter-stream synchronization control after carrying out the inter-stream synchronization control for the slave streams.

In remote cooperation, in order to improve the efficiency of cooperative work, it is important to output MUs simultaneously at different terminals. Group (or inter-destination) synchronization control outputs each MU simultaneously at different terminals. We proposed three schemes for inter-destination synchronization control (i.e., the master–slave destination scheme, synchronization maestro scheme, and distributed control scheme) [4].

5.2 Causality control

Causality control keeps the causal (i.e., temporal order) relationships among events. Here we introduce two typical examples of causality control; one is the Δ-causality control [20], and the other is the adaptive Δ-causality control [21].

In the Δ-causality control, each MU has a time limit which is equal to the generation time of the MU plus Δ seconds for preservation of the real-time property. The control output the MU at the time limit, and if the MU is received after the time limit, it is discarded because it is considered useless. The adaptive Δ-causality control dynamically changes the value of Δ according to the network load. The control does not discard an MU received after the time limit and uses the MU for prediction.

5.3 Adaptive reaction force control

As the network delay increases, the reaction force applied to a haptic interface device becomes larger and the output quality of haptic media becomes deteriorated. The adaptive reaction force control [4] can be used to solve the problem. We calculate the reaction force based on the spring-damper model [22] or depending on the force sensed by the force sensor. In the spring-damper model, the reaction force consists of the elasticity and viscosity. The elasticity is force exerted by deformation of a spring or rubber, for example. When a spring is pushed or pulled. The elasticity is proportional to the depth of a spring when the spring is pushed, and it is calculated by multiplying the depth by the elastic coefficient. The viscosity is force or resistance exerted by fluids, for example, when we move an object through the fluids (e.g., water and oil). The viscosity is proportional to the relative velocity (i.e., the velocity of the object relative to the fluids), and it can be calculated by multiplying the relative velocity by the viscosity coefficient. The adaptive reaction force control includes the adaptive viscosity control [23], adaptive elastic control [24], and adaptive viscoelasticity control [25]. The adaptive elastic control dynamically changes the elastic coefficient according to network delay, the adaptive viscosity control dynamic changes the viscosity coefficient according to the network delay and the velocity of the haptic interface device, and the adaptive viscoelasticity control combines the two types of control.

5.4 Position control using force information

In order to reduce the force applied to an object operated in cooperative work between the remote robot systems with force feedback, the robot position control with using force information is proposed [17]. The proposed control moves the robot by taking advantage of human perception of force direction by experiment. The control finely adjusts the robot position dynamically in the direction where the force is reduced.

Since the remote robot system with force feedback is delay sensitive [26], we apply Skipping to the system at both master and salve terminals. This means that Skipping is applied to the operation with a single remote robot system and that between the two remote robot systems. For the operation between the two remote robot systems, we apply the adaptive reaction force control, adaptive Δ-causality control, and position control using force information for the remote cooperation between two remote robot systems with force feedback. That is, we apply the control for the cooperation between users (i.e., each user operates a haptic interface device to control a remote robot to do collaborative work), and for the cooperation between the user and robot. We also applied the position control using force information for the cooperation between the two robots. We also investigate the effects of the control by experiment.

6. Challenges and future directions of QoS control

As described in the previous section, although we proposed serval types of QoS control for the remote robot system with force feedback, there still exist many challenges. In this section, we discuss the challenges and future directions of QoS control.

6.1 Integration of QoS control and stabilization control

In order to achieve stable and high quality of service, we also need to carry out stabilization control, and it is important to integrate the QoS control with stabilization control. To integrate the QoS control with stabilization control, we can carry out QoS control and stabilization control independently, or integrate QoS control into stabilization control for the system. We investigate the effects by integrating the position control using force information as QoS control with stabilization control with filters [9], and experimental results show that the effect when we carry out QoS control in the loop of stabilization control is better than that when we carry out QoS control and stabilization control independently. It is important to investigate the effect by using other types of QoS control and stabilization control.

6.2 Multilateral control

In this chapter, we introduced the QoS control only in a communication loop, which is between a haptic interface device and a remote robot (see **Figure 3**). However, in the remote operation using multiple systems, there are multiple loops caused by communication in the systems (see **Figure 4**), and there exist inter-relationships among the loops. This means that we need to carry out multilateral control for QoS as well as stabilizations and it becomes complex and difficult.

6.3 Application of big data, cloud computing and AI technologies

In order to improve the efficiency of QoS control, we need to take account of many factors, for example, contents of work, movement speed, room temperature and wind [27]. Therefore, big data [28], cloud computing [29], and AI (Artificial Intelligence) [30] technologies such as neural network, fuzzy theory, and genetic algorithm can be useful methods for efficient control. The necessary information for QoS control can be transmitted to a cloud server, and the information can be combined as big data for analysis and applied as training data and evaluation data. Efficient QoS control can be expected by using AI after studying the training data. Also, in order to solve the problem of AI computing, we can apply AI chips [31],

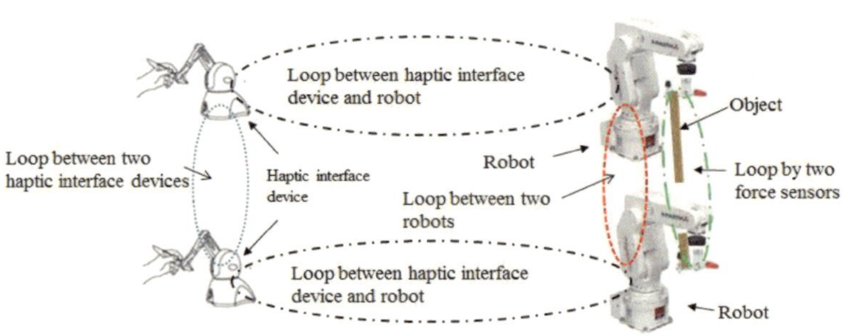

Figure 4.
Control loops in remote robot operations.

which realizes edge AI computing, to the remote robot terminal to improve the efficiency of QoS control.

6.4 Others

Since we need to transmit the necessary information for QoS control to a cloud server, it is important to consider the safety and security of data. Also, in many situations, we need to use movable robots as remote robots. This means that we may need to consider the QoS control in wireless and/or mobile networks. This is because a 5G network [32] which is wideband and low latency becomes available and the possibility of the application over the mobile network increases.

In addition, we need to carry out QoE assessment to investigate the effects of QoS control and to clarify how to set parameter values optimally under each type of the control as well as QoS assessment at lower layers. QoE subjective assessment is the most important because the assessment can reflect end users' opinions directly [4], [33–35].

7. Conclusions

In this chapter, we focus on QoS control for remote robot operation. We introduce our remote robot system with force feedback which we constructed to study QoS control. We also present the expected applications and the problems to be solved for widespread application of remote robot system with force feedback. We mainly focused on the problems of network delay, delay jitter, and packet loss. We explain several types of QoS control which we previously proposed to solve the problems. Finally, we also discuss the challenges and future directions of QoS control.

For the future plan of our study, we need to solve the problems described in Section 6.

Acknowledgements

The authors thank Prof. Hitoshi Watanabe of Tokyo University of Science, Prof. Hitoshi Ohnishi of the Open University of Japan, Prof. Takanori Miyoshi of Nagaoka University of Technology, and Prof. Takashi Okuda of Aichi Prefectural University for their valuable discussions.

Conflict of interest

The authors declare no conflicts of interest associated with this chapter.

Author details

Pingguo Huang[1*] and Yutaka Ishibashi[2]

1 Gifu Shotoku Gakuen University, Gifu, Japan

2 Nagoya Institute of Technology, Nagoya, Japan

*Address all correspondence to: huangpg@gifu.shotoku.ac.jp

IntechOpen

© 2021 The Author(s). Licensee IntechOpen. This chapter is distributed under the terms of the Creative Commons Attribution License (http://creativecommons.org/licenses/by/3.0), which permits unrestricted use, distribution, and reproduction in any medium, provided the original work is properly cited.

References

[1] K. Ohnishi , Real world haptics: Its principle and future prospects. The Journal of the Institute of Electrical Engineers of Japan, 2013. 133, 268-269. (In Japanese) https://doi.org/10.1541/ieejjournal.133.268.

[2] T. Kawai, Haptics for surgery. The Journal of the Institute of Electrical Engineers of Japan, 2013. 133, 282-285. ttps://doi.org/10.1541/ieejjournal.133.282

[3] Y. Ishibashi and P. Huang, Improvement of QoS in haptic communication and its future. The IEICE Transactions on Communications, 2016. J99-B, 1911-925. (Japanese Edition).

[4] P. Huang and Y. Ishibashi, QoS control and QoE assessment in multi-sensory communications with haptics. The IEICE Transactions on Communications, 2013. E96-B, 392-403. https://doi.org/10.1587/transcom.E96.B.392.

[5] ITU-T Rec. I. 350, General aspects of quality of service and network performance in digital networks. 1993.

[6] ITU-T Rec. G. 100/P. 10 Amendment 1, New appendix I - Definition of quality of experience (QoE). Jan. 2007.

[7] T. Miyoshi, Y. Maeda, Y. Morita, Y. Ishibashi, and K. Terashima, Development of haptic network game based on multi-lateral tele-control theory and influence of network delay on QoE. Transactions of the Virtual Reality Society of Japan (VRSJ), Special Issues on Haptic Contents, 2014. 19, 559-569. (In Japanese)

[8] P. Huang, T. Miyoshi, and Y. Ishibashi, Stability control in remote bilateral robot control system with force feedback. Proc. IEEE The 3rd International Conference on Control, Automation and Robotics (ICCAR), pp. 22-24 April 2017.

[9] P. Huang, T. Miyoshi, and Y. Ishibashi, Enhancement of stabilization control in remote robot system with force feedback. International Journal of Communications, Network and System Sciences (IJCNS), vol. 12, no. 7, pp. 99-111, July 2019.

[10] https://www.3dsystems.com/haptics-devices/touch

[11] http://www.mitsubishielectric.co.jp/fa/products/rbt/robot/lineup/manual/f/bfp-a8899k.pdf (In Japanese)

[12] http://dl.mitsubishielectric.co.jp/dl/fa/members/document/manual/robot /bfp-a8940/bfp-a8940Z.pdf (In Japanese).

[13] K. Suzuki, Y. Maeda, Y. Ishibashi, and N. Fukushima, Improvement of operability in remote robot control with force feedback. Proc. The 4th IEEE Global Conference on Consumer Electronics (GCCE), pp. 16-20, Oct. 2015.

[14] http://dl.mitsubishielectric.co.jp/dl/fa/members/document/manual/robot/bfp-a8080/ bfp-a8080e.pdf (In Japanese).

[15] P. Huang and Y. Ishibashi, QoE assessment of will transmission using vision and haptics in networked virtual environment. International Journal of Communications, Network and System Sciences (IJCNS), vol. 7, no. 8, pp. 265-278, Aug. 2014.

[16] P. Huang, Y. Ishibashi, and T. Miyoshi, Stabilization in remote robot systems with force feedback. (In Japanese), IEICE Technical Report, CQ2020-97, Jan. 2021.

[17] Y. Ishibashi, E. Taguchi, P. Huang, and Y. Tateiwa, Robot position control

with force information in cooperation between remote robot systems. Proc. The 5th International Conference on Control, Automation and Robotics (ICCAR), pp. 147-151, Apr. 2019.

[18] Y. Ishibashi, A. Tsuji, and S. Tasaka, A group synchronization mechanism for stored media in multicast communications. Proc. 16th IEEE International Conference on Computer and Communications (INFOCOM), pp.693-701, April 1997.

[19] Y. Ishibashi, S. Tasaka, and T. Hasegawa, The virtual-time rendering algorithm for haptic media synchronization in networked virtual environments. Proc. the 16th International Workshop on Communications Quality and Reliability (CQR), pp.213-217, May 2002.

[20] R. Yavatkar, MCP: A protocol for coordination and temporal synchronization in multimedia collaborative applications. Proc. The 12th International Conference on Distributed Computing System (ICDCS), pp.606-613, June 1992.

[21] Y. Ishibashi and S. Tasaka, Causality and media synchronization control for networked multimedia games: Centralized versus distributed. Proc. The 2nd Workshop on Network and System Support for Games (NetGames), pp.34-43, May 2003.

[22] 3D Systems, OpenHaptics toolkit programmer's guide. version 3.2, 2013.

[23] Y. Komatsu, H. Ohnishi, and Y. Ishibashi, Adaptive control of viscosity in remote control system with force feedback. Proc. IEEE International Conference on Consumer Electronics - Taiwan (ICCE-TW), pp. 237-238, June 2017.

[24] S. Suzuki, K. Matsunaga, H. Ohnishi, and Y. Ishibashi, Influences of network delay variation on haptic perception under adaptive reaction force control. Proc. The 3rd IEEE Global Conference on Consumer Electronics (GCCE), pp. 669-673, Oct. 2014.

[25] T. Abe, Y. Komatsu, H. Onishi, and Y. Ishibashi, QoE assessment of adaptive viscoelasticity control in remote control system with haptic and visual senses. Proc. IEEE International Conference on Consumer Electronics - Taiwan (ICCE-TW), pp. 133-134, May 2018.

[26] S. Matsumoto, I. Fukuda, H. Morino, K. hikichi, K. Sezaki, and Y. Yasuda, The influence of network issues on haptic collaboration in share virtual environments. Proc. the 5th PHANToM User Group Workshop, Oct. 2000.

[27] X. Wang, P. Huang, Y. Ishibashi, T. Okuda, and H. Watanabe, Influence of network delay on QoS control using neural network in remote robot systems with force feedback. Proc. 2020 The 9th International Conference on Networks, Communication and Computing (ICNCC), Dec. 2020.

[28] I. A. T. Hashem, I. Yaqoob, N. B. Anuar, S. Mokhtar, A. Gani, and S. U. Khan, The rise of 'big data' on cloud computing: Review and open research issues. Information Systems, vol. 47, pp. 98-115, Jan. 2015.

[29] V. A. Memos, K. E. Psannis, Y. Ishibashi, B. G. Kim, and B. B. Gupta, An efficient algorithm for media-based surveillance system (EAMSuS) in IoT smart city framework. Future Generation Computer System, vol. 83, pp. 619-628, 2018.

[30] A. Ramesh, C. Kambhampati, J. Monson and P. Drew, Artificial intelligence in medicine. Annals of The Royal College of Surgeons of England, vol. 86, pp. 334-338, Oct. 2004.

[31] H. Fuketa and K. Uchiyama, Edge artificial intelligence chips for the cyberphysical systems era. IEEE

Computer, vol. 54, no. 1, pp. 84-88, Jan. 2021.

[32] M. Shafi, A. F. Molisch, P. J. Smith, T. Haustein, P. Zhu, P. D. Silva, F. Tufvesson, A. Benjebbour, and G. Wunder, 5G: A tutorial overview of standards, trials, challenges, deployment, and practice. *IEEE Journal on Selected Areas in Communications*, vol. 35, no. 6, pp. 1201-1221, June 2017, doi: 10.1109/JSAC.2017.2692307.

[33] Q. Qian, Y. Ishibashi, P. Huang, Y. Tateiwa, H. Watanabe, and K. E. Psannis, Softness comparison of stabilization control in remote robot system with force feedback. Proc. IEEE TENCON, pp. 32-37, Oct. 2018.

[34] Q. Qian, D. Osada, Y. Ishibashi, P. Huang, and Y. Tateiwa, Human perception of force in cooperation between remote robot systems with force feedback. International Journal of Mechanical Engineering and Robotics Research (IJMERR), vol. 9, no. 2, pp. 264-269, Feb. 2020.

[35] L. Wen, Y. Ishibashi, P. Huang, Y. Tateiwa, and H. Ohnishi, QoE assessment of weight perception in remote robot system with force feedback. Proc. The 2nd World Symposium on Communication Engineering (WSCE), pp. 200-204, Dec. 2019.

Chapter 5

Using Ontologies in Autonomous Robots Engineering

Esther Aguado and Ricardo Sanz

Abstract

The construction and operation of autonomous robots is heavily based of systemic conceptualizations of the reality constituted by the robot, its controller and the environment where it performs. In this chapter we address the role that computer ontologies play in the whole life cycle—engineering and operation—of autonomous robots: from its conception and construction by human engineering teams to deployment and autonomous operation in dynamic and uncertain environments. This chapter summarizes the state of the art, gives some examples and establishes a roadmap for future activity in this domain to produce shareable ontologies that could streamline autonomous robot development and exploitation.

Keywords: Robotics, Engineering, Ontology, Autonomy, Adaptation

1. Introduction

Technical systems are designed and built to perform a variety of operations in pursue of user needs. In many cases we want these systems to operate without human intervention for performance, cost or safety reasons. We want many technical systems to be autonomous. From fridge thermostats to the country-wide electrical utilities we expect 100% availability even in changing circumstances.

Autonomy requires the ability to perform the assigned task without external help. The autonomous car shall be able to negotiate an intersection and the autonomous space probe shall be capable of reorienting itself. But autonomy without robustness is a no go. Real autonomy also requires capabilities for enduring disturbances during operation. Conventional control systems are built to overcome some forms of disturbance—up to a limit. Autonomy shall be robust to operate in a wider range of circumstances.

In many cases, these systems need to be endowed with autonomy features to be able to operate in unstructured and hazardous environments. Many real-world situations show high levels of disturbance and uncertainty that displace the system from the normal operating region for which it was designed. In other cases, the disturbances come from inside the system. Electrical interference or device faults can lead the system to mission-level failure.

In all these situations—when the system is pushed out of the designed region of operation—the system requires certain adaptation capabilities to provide the levels of robustness and resilience necessary to overcome the adverse situation and ensure mission fulfillment.

As said, these disturbances may come from outside the system, such as changes in the environmental conditions, *e.g.* amount of obstacles, terrain characteristics, *etc.*, or from the system itself, such as failures in individual components used to perform a specific task or providing communication mechanisms. To overcome localized failures, autonomous systems require, when deployed, a certain level of redundancy to enable the use of fault-handling techniques to overcome these disturbances. These redundancies can be structural—e,g, when having spare components or using triple modular redundancy—or functional—*e.g.* when having different ways of reducing car speed.

In any case, all these mechanisms require ways of representing the knowledge about them both at design time—by engineers building the system—and at run-time—by the autonomous system itself. Engineers capture this knowledge in the engineering models and autonomous systems may use it in knowledge-driven perception-decision-action loops.

In this chapter, we address the use of ontologies as substratal assets for these knowledge based processes. Ontologies can be used to decouple those conceptual elements for system adaptation from the particular implementation used in a concrete deployed system. Decoupling design knowledge and realization promotes reusability, modularity, and scalability. All of them, critical properties of sound engineering processes. Ontologies can provide a shared understanding of all stages of the system life cycle—from conceptualization to decommission—to both ease the task of the engineer and improve system run-time operation. In this chapter, we specifically focus on the benefits of using ontologies in autonomous systems, especially in autonomous robots, and present an implementation case with adaptive mobile robots.

The chapter is organized as follows: Section 2 defines what is an ontology and collects well-known ontologies for autonomous robots. Section 3 defines what is autonomy and other the key concepts we aim to reach in this type of systems. Section 4 presents the scope of ontologies within the life cycle of autonomous systems. Section 5 addresses a concrete proof-of-concept of ontologies for augmenting the autonomy level in a mobile robot. Section 6 discusses the general implications at system level when using ontologies. Section 7 presents a roadmap for the use of ontologies to streamline autonomous robot development and exploitation. Lastly, Section 8 presents the conclusions of the chapter.

2. Ontologies for autonomous robots

Ontology—with upper case—is the branch of philosophy dedicated to the study of being. From this perspective of analyzing what exists, derives the use of ontologies—with lower case—in computer and information science. Computer ontologies are specifications of conceptualizations [1]. They formally document the types of entities that exist in a domain, their properties, and the relationships between them[1]. A conceptualization is an abstract, simplified perspective in some area of interest. The conceptualization includes the objects, concepts, and other entities and the relationships among them. Ontologies are used in information systems to guarantee the conformance of a knowledge base with a certain conceptual specification.

Ontologies are built on top of terms that are used to capture the concepts. They provide a formal naming and definition of categories, properties, and relations between concepts, data, and entities. In practice, ontologies are just computer-readable files in a specific computer language that reify a conceptualization of elements of a specific domain.

[1] To be precise, they formally document *the information about* the types of entities, their properties, and the relationships between them.

In general, ontologies define a common vocabulary for a domain, allowing the reuse of domain-specific knowledge. Moreover, it provides a common understanding about a field for both people and artificial agents. It is this shareable knowledge among humans and software entities that makes ontologies a valuable asset in autonomous robot engineering. In [2] the authors remark the importance of formally represented knowledge and its foundation in *conceptualizations* that are shared among people. In our case these conceptualizations will be shared among humans—engineers—and intelligent machines—robots. These concepts are widely used in computational operations nowadays. For instance, in [3], the authors use an ontology-based method to assign tasks to a satellite cluster; in [4] they use ontology-aided reconfiguration to manage an IT infrastructure; and in robotics, works like the one presented in [5], make use of ontologies to define an information model to integrate the knowledge of heterogeneous fleets of robots in underwater operations.

2.1 Ontologies in robotics

The IEEE 1872-2015 standard [6] was published in early 2015. This is a standard for the robotics and automation domain that defines a set of ontologies for robots. The most popular of these ontologies is the core ontology for robotics and automation (CORA) which specifies the most general concepts in the robotics domain [7].

This standard is based on the Suggested Upper Merged Ontology (SUMO) [8]. SUMO—a top-level ontology—is used to provide differentiation among terms that refer to physical and abstract entities and serves as the basis of the robotics ontology. In SUMO, classes such as *agent*, *process* or *proposition* are defined. This knowledge is then specialized by CORA with classes such as *Robot* or *Robotic System*. Other ontologies defined in this standard are the position (POS) ontology [9] that captures the main concepts and relations regarding the position, orientation, and pose which are key elements for robot navigation and manipulation, and the RPARTS ontology that provides a set of specific types of roles that specialize the general role of robot parts.

The IEEE 1872 ontologies are explicit and formal, however, they are maybe too general for practical use. For this reason, there are a variety of more specific ontologies with different scopes in the robotics domain. Olszewska [10] presents an autonomous robot architecture ontology (ROA). ROA defines the main concepts and relations for defining a robot architecture. This ontology has been tested in driving a human-robot interaction scenario. A human operator specifies a task and this task is divided and associated with a robot according to its capabilities. All these types of entities are captured in the ontology and used to create information elements that enable human-robot communication. In [11] an extension of CORA is presented with concepts of design, environment, and interaction on artificial systems.

Most of these ontologies reuse knowledge from other ontologies. Sometimes ontologies are identified with knowledge bases, especially after the use of standardized languages to capture both generic and specific knowledge. In order to facilitate reuse, ontologies are designed following the principle of modularisation. Modularity provides a series of benefits in contrast to the problems of using big, complex, monolithic ontologies [12]:

- Scalability for querying data and reasoning. Small-scale ontologies allow easier concept assertion and reasoning than when handling a large number of entities.
- Scalability for update and maintenance. Ontologies, as any other artifact requires maintenance and may need to be updated and enlarged with new knowledge. This task is easier if the knowledge is structured in modules.

- Complexity management. The design process is more straightforward when working with small modules—*i.e.* a reduced set of concepts—integrated into the final ontology.
- Understandability. It is easier to understand an ontology in small portions than a huge ontology, either in a textual or visual form.
- Context-awareness. The use of modular ontologies simplifies the contextualization in the creation of knowledge. Each module can focus on any aspect regarding one context.
- Reusability. The split of an ontology into modules provides the reuse of specific parts in other ontologies.

Nevertheless, modular ontologies may lead to some problems during the process of creation and use. For example, the integration of other concepts by importing existing ontologies may lead to unexpected consequences such as inconsistencies related to reused vocabulary—*e.g.* conflicting definitions for homonyms. Keeping safety and correctness in a modular ontology is a key element when extracting or importing knowledge among ontologies and modules.

2.2 Standard ontologies beyond CORA

As said, CORA is a quite general standard that is not very effective for concrete applications. The above-mentioned ontologies by Olszewska or Fiorini try to provide more specific ontologies directly usable in concrete applications.

This fact is recognized by the Standards Association of the IEEE who is currently developing a collection of CORA-based standards to address different aspects of the robotics domain. These standards address different aspects of relevance like task specification, autonomy, ethics, agility, or verification of autonomous behavior.

In particular, the Robotics and Automation Society Standing Committee on standards working group 1872.2[2] is elaborating a CORA-based standard ontology for autonomous robotics [13]—the Autonomous Robotics (AuR) Ontology.

The AuR standard under development shall extend the CORA ontology by defining additional ontologies for the autonomous robots domain. These ontologies will address different aspects of relevance: (i) general concepts for autonomous robots; (ii) core design patterns specific to autonomous robot systems; and (iii) general use cases and/or case studies.

2.3 Ontologies and model-driven engineering

Currently, there are no generally accepted method or framework for the design of complex robotic systems [14]. However, this task of building complex robotic systems can easily leverage extant systems and software development methods. In complex system developments, the design is focused at a different level of abstractions, and modularity is used to both organize the design and implement the system [15]. Examples of this modular approach are the developments based on object-oriented methods, middleware, and component-based design.

The structural, modular organization of design knowledge and the exploitation of formally captured system knowledge is the basement of the large collection of

[2] https://standards.ieee.org/project/1872_2.html

model-based approaches in model-driven engineering (MDE). For example, OMG's Model-driven Architecture (MDA) focuses on design models a level of abstraction up of objects and components to reach modular reusable abstractions that can be later particularised for specific uses (Platform-Independent Model (PIM) → Platform-Specific Models (PSM)). In the system's domain—closer to the robotics domain—languages like Systems Modeling Language (SysML) are gaining momentum due to their universality as a vehicle for augmented design formalization. However, MDE methods often suffer from a lack of semantics and truly formal knowledge representations that can be effectively exercised [16].

To overcome these bottlenecks, a formal ontology can be included as part of the model definition. One example of a model that makes use of ontologies to specify the system behavior and architecture is the Teleological and Ontological Model for Autonomous Systems (TOMASys) framework [17]. TOMASys is a domain-independent metamodel that allows the construction of models to define architectural alternatives in component-based systems. This metamodel is teleological because it incorporates core concepts in the engineering conceptualization as are the concepts of system intention and the purpose of the designers when creating a specific subsystem. And it is ontological because it defines a formal vocabulary for systems structure and behavior.

The ontological approach followed in this chapter and presented in the proof-of-concept in Section 5 is built upon the TOMASys framework that was designed following the ideas of model-based systems engineering and particularized for the autonomous system engineering domain.

3. Autonomy and relate

The work described in this article addresses the use of ontologies for the augmentation of autonomy in robots [18]. As ontologies foster the use of formality in conceptualizations, it seems natural to try to provide a definition of the adjective *autonomous*.

The term "autonomous" is a buzzword these days and has received different meanings in different contexts[3]. In the analysis of the use of the term "autonomous" in automatic control and robotics, there are two major generalized uses of the term "autonomous":

- A robot is said to be autonomous if it has the capability of *moving* by its own resources and under self-control.

- A robot is said to be autonomous if it has the capability of performing certain tasks without human—or external—help. A task-generalization of the former.

In our position as autonomous *systems* engineers, it is the second interpretation that we focus on. In systems engineering the task to be performed by the system is always something of value to the final user. An *useful* mobile robot shall not just wander around but perform some task of value during this wandering (find an object, move an object, detect intruders, *etc.*). When we say that a robot is autonomous we mean that it is capable of performing its assigned activities—*e.g.* generate a map—without the need of external intervention [19]. This also applies to the

[3] It has indeed a long tradition of use in the domains of healthcare and political science.

capability of movement—including the whole robot navigation infrastructure—and to all the other functions that the robot may perform subsidiarily to the main task [20].

3.1 Autonomy and disruption

A second aspect concerning task execution that is of maximal importance is the distinction between (i) being able to perform certain tasks alone (*e.g.* moving to a pose or building a map); (ii) doing so while handling some degree of disturbance; and (iii) being able to perform these tasks *alone* in the presence of *severe disturbances*[4]. In the first case, a simple automaton can do the job. In the second case, a feedback control system can do the job. In the third case, a perception-thought-action loop is necessary to provide both feedback, adaptation, and anticipation. Some people use the term "automatic" for the first or second cases, keeping "autonomous" for the third. In the automatic control domain, some authors may use "open loop" and "closed-loop" to make this distinction, but for us, the second case also includes closed-loop controllers for operational set-points.

A more thorough distinction could be done concerning the nature of the disturbances, especially when severe. In the case of anticipated, well known severe disturbances, the system could be built in accordance to them to be able to respond adequately and predictably. If the disturbances are not predictable—or don't want to bother about their anticipation—the system can be built to respond reactively to them. In the design of the system, we shall define, however, a set of bounds of the system operational environment to be able to design the system to behave robustly in this region.

In the work described in this chapter, we address situations where the system finds itself outside the boundaries set for its operation at design time—its normal operational profile. In these circumstances, the only possibility for keeping the mission going is for the robot to adapt to the new situation: it shall change its very design/realization to be able to still achieve mission objectives in this new situation.

3.2 Autonomy and trustworthiness

Trustworthiness is a necessary but not sufficient condition to carry out tasks in open environments [23]. In real operation, autonomous systems are deployed in complex environments plagued with uncertainty. This affects the system capability to complete the mission assigned to it by the user. For a user to confidently rely on an autonomous system, the system shall be trustworthy.

Trust and trustworthiness may seem similar but they must be distinguished; especially in an autonomous system, where behavior assurance is quite more complex. Trust is a human-system relational property; *i.e.* something that the human user perceives or *feels about* the robot. On the contrary, trustworthiness is a property of the system itself, *i.e.* that the system is robust and resilient in relation to its mission and hence, it deserves trust by the human user [24]. This implies that a human user may not trust a trustworthy system [25] because user perception is

[4] A severe disturbance is a disturbance that violates the system design assumptions for normal operational conditions. An example of severe external disturbance is a slippery floor for an unmanned ground vehicle (UGV) when designed to operate on a non-slippery floor. An example of severe internal disturbance is the failure of a laser range sensor used in robot navigation. See [21] for a discussion of types of system change under the Klir general systems framework [22].

affected by limited knowledge, observation capability, and biased by previous experiences.

The problem we are addressing here is achieving trustworthiness, specifically dependability and mission assurance. The framework discussed here provides engineering tools in terms of system and mission concepts and relationships to define system design alternatives to deal with abnormal scenarios and unpredictable environments.

The underlying idea is to break the design/operation barrier. Using ontologies we can make available the engineering design knowledge at run-time to allow system self-reconfiguration using self-knowledge. With this approach, the scope of the ontologies covers from the system conceptualization until the system deployment. The use of ontologies at run-time provides an information-driven adaptation capability to enhance system autonomy [18].

4. Ontologies in the life cycle of autonomous systems

In systems engineering, the life cycle of an artifact usually includes eight stages: (i) identify the needs, (ii) define the system concept, (iii) specify system requirements, (iv) design the system, (v) implement the system, (vi) verify the system, (vii) deploy the system and (viii) operate it[5]. In the first six stages, the work is typically iterative until the deployment phase, when requirements and design decisions are frozen and remain implicit in the final artifact.

In fault-tolerant systems, a set of methods and algorithms intervene at run-time to keep the functional activity of the system, *i.e.* to maintain the operation as it was designed. The fault-tolerance mechanics is predefined, blind, and triggered by certain events. There is no system knowledge to reason about but its reification in rigid adaptation mechanisms. The idea we pursue in this work is the usage of ontologies to include the knowledge of engineering as part of the run-time system to endow the system with *flexible reconfiguration capability based on system knowledge*. With this approach, the design phase and the deployed phase maintain an explicit link through the system knowledge because the system ontology provides a metamodel that spans the whole system life cycle. This link can be exploited to combine other subsystems and create new designs at run-time more suitable for addressing certain contingencies.

Ideally, the system knowledge base should encode all include all the system concepts developed in early phases of the system life cycle, for example, *user needs* as the artifact is produced to satisfy the needs defined in the first stage. With this information, the system could be able to ensure the mission and reason about it at any stage.

In adaptive systems, with component or functional redundancy, the early stages of the life cycle are not addressed. The reconfiguration in this case aims to comply with the initial system design or a few designs for possible known contingencies.

However, by providing the system with capabilities to trace until the needs that justify its existence as well as the requirements that justify that design, the system can augment its autonomy in search of trustworthiness. If a requirement is imposed by a component that is not functioning and is going to be substituted with another element, that requirement is no longer applicable to the system. Therefore, besides the component in use, other adjustments can be made in the system for better performance.

[5] A final decommissioning stage is also of importance, esp. in terms of sustainability, for real-world systems. We do not address this stage here.

An example of this case is the use of different navigation sensors in a mobile robot. Suppose we have an autonomous robot with laser and ultrasound sensors to navigate. An initial objective may be to reach a point as fast as possible. According to the final design of the robot, that requirement would be specified with a specification of a targeted velocity value.

The laser is a device with a high refresh rate so the robot can navigate safely at higher velocities. If the robot enters a room with glass walls, the laser is not reliable. If the robot detects through the reasoning that the environmental conditions are not suitable for the laser and triggers a reconfiguration to use the ultrasound sensor, the robot can keep its operation to fulfill the mission. However, as the design has changed, the requirements that can be fulfilled are not the same. In this case, as the ultrasound sensor has a shorter range, the maximum velocity of the robot shall be significantly less to keep a safe operational profile. Once the robot has traversed that glass room, the laser can be re-activated so the requirements must change again to achieve the maximum performance available.

This is a naive example of how a system engineering knowledge base can improve a navigation task. However, real-world missions are composed of complex-orchestrated tasks, for instance, the operation of a waiter-robot which must serve a drink, or a miner-robot that must obtain a certain mineral. In this case, that knowledge can be further exploited with deep reasoning to perform adaptation at different tasks and several stages of the system life cycle.

5. Fault-tolerant mobile robot proof-of-concept

Following the naive example above, an ontology-driven reconfiguration capability for mobile robots has been explored in the Metacontrol for Robot Operating System (MROS) project[6]. In this implementation, the robot's mission is to move to a certain point. During the mission, several contingencies may occur. In this case, we contemplate the internal contingency cases of (i) laser rangefinder failure and (ii) low battery. Additionally, the mission has some operational requirements associated in terms of performance, safety, and energy consumption that the robot must ensure during the navigation.

The assurance of the operational requirements along with the contingency handling is governed by a knowledge base structured on description logics and exerted by a reasoner. The key of this approach is the usage of the general modeling framework presented in Section 2.3, TOMASys. The TOMASys ontology is particularized with two sets of individuals: the navigation-domain ontology and an application-specific ontology. We use a modular approach in the construction of the ontology to be capable of reusing a part of the non-specific knowledge in any navigation application in mobile robotics. The ontologies are instantiated for run-time use as a knowledge base composed of three OWL 2 [26] files.

The TOMASys metamodel is used to depict structure and behavior with an explicit representation of the objectives of the system as well as the components required to realize them. The system concepts provided by this metamodel are divided into two main groups:

- The static knowledge is stored in `Functions` and `Function Designs`. The `Function` element allows the definition of abstract `Objectives` for the system to complete the mission. The `Function Design` element stores all the design

[6] https://robmosys.eu/mros/

alternatives the system engineer has thought as possible to fulfill a certain Function.

- The instantaneous state is captured with Objectives, which define a set of operational requirements pursued at run-time when executing a Function; Function Groundings, that are used at run-time to specify which Function Design is in use; and Components, used to describe the structural modules at that instant. Lastly, Quality Attributes affect both static and run-time knowledge. They are used to make explicit the operational requirements of the mission.

- Each Objective has a Quality Attribute associated to meet operational requirements such as safety, performance, and energy consumption. Likewise, each Function Design has a Quality Attribute value estimation to select the best design alternative to meet the mission requirements. Additionally, the Function Grounding measures the real Quality Attribute value to monitor if those requirements are being fulfilled.

As it was previously mentioned, the knowledge base is completed with two sets of individuals. The navigation-domain file contains instances of widely-used navigation sensors such as ultrasound, laser, RGBD cameras, etc., and other important elements in autonomous robots such as the battery. These elements are instances of the TOMASys class Component. Besides, popular Quality Attributes are defined such as energy, safety, and performance.

The application-specific knowledge base is made of all the Function Designs, these are the design alternatives to perform navigation. Other elements are the instance of an Objective, the instance of a Function Grounding, this is the Function Design in use, and the Quality Attributes relative to them. Each Function Design has a Quality Attribute estimation in safety and energy, which is calculated for the Function Grounding. This calculated Quality Attribute value is compared with the non-functional requirements (NFR) defined for the Objective. The NFRs are the Quality Attributes required for the specific mission.

5.1 Run-time reconfiguration for fault-tolerance

To use the knowledge base at run-time, it is written in a machine-readable format using the Web Ontology Language (OWL). A descriptive logic (DL) reasoner uses it during the system operation to evaluate the robot's functioning. Once an Objective is defined, and it is linked to the Function that solves it, a Function Grounding is selected according to the mission requirements and the Component availability. In the MROS proof-of-concept, two possible classes of contingencies are addressed: component fault and mission requirements non-fulfillment.

Each Component has a *required by* relationship with the Function Design that makes use of it. If a Component is malfunctioning, those Function Designs that use it becomes unavailable. **Figure 1** depicts the main relationships contained in the knowledge base. The two components considered, laser and battery, are required for all Function Designs except one. In case of laser failure, the Function Design degraded mode should be selected. Likewise, in the case of a low battery, the Function Design energy saving mode should be selected. This is implicitly shown in the figure, as there are no links between those Function Design individuals and the corresponding Component individual.

The ontology also includes some rules using the Semantic Web Rule Language (SWRL) to perform functional diagnosis. This is done by asserting the information

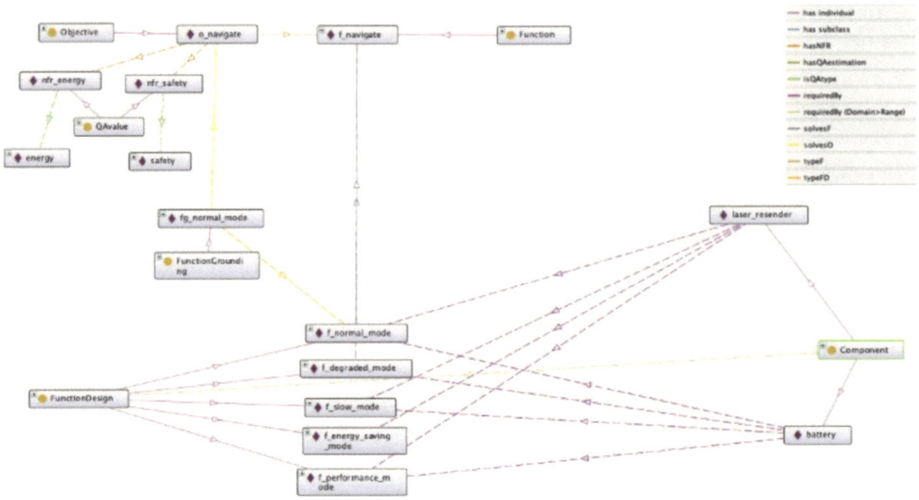

Figure 1.
Main individuals and relationships of the proof-of-concept knowledge base. The Objective o_navigate is fulfilled by the Function Grounding fg_normal_mode, this Function Grounding is a realization of the Function Design f_normal_mode which solves the Function f_navigate. This Function is required as is the one that solves the Objective. Component required for this Function Design are laser and battery. Among all the possible Function Design, the one that does not require the laser if it becomes unavailable is f_degraded_mode. Additionally, Quality Attributes values relative to the non-functional requirements (NFRs) of the Objective are depicted for the Quality Attributes of safety and energy.

about the status of the Components that compose the system, the design in use to solve a function (Function Grounding), and the status of the Objective.

There are three sets of rules that: (i) set the Objective in error if the objective requirements, NFR Quality Attributes, are not met; (ii) set an Function Grounding in error if a Component in use is in error; and (iii) propagate Function Grounding error to the Objective. Lastly, there are some additional rules regarding the storage in a log file of the Function Grounding that have been in error and the status of a Function Design realisability depending on the status of the Components. For instance, if the laser is in error, the only Function Design with realisability with a true value will be *f_degraded_mode* according to **Figure 1**. **Table 1** shows three example SWRL rules used in this proof-of-concept.

Rule no.1 *tomasys:Component(?c) tomasys:c_status(?c, false) mros:requiredBy(?c, ?fd) tomasys:typeFD(?fg, ?fd) tomasys:FunctionGrounding(?fg) → tomasys:fg_status(?fg, INTERNAL_ERROR)*
If a Component has a status in false (in error), and that component is *required by* a Function Design with the same *type* as the Function Design in use, Function Grounding, then that Function Grounding status is set as *INTERNAL ERROR*.
Rule no.2 *tomasys:FunctionGrounding(?fg) tomasys:fg_status(?fg, INTERNAL_ERROR) tomasys:solvesO(?fg, ?o) tomasys:Objective(?o) → tomasys:o_status(?o, INTERNAL_ERROR)*
If a Function Grounding has a status in *INTERNAL ERROR*, and that Function Grounding *solves* an Objective, then that Objective status is set as *INTERNAL ERROR*.
Rule no.3 *tomasys:Component(?c) tomasys:c_status(?c, false) mros:requiredBy(?c, ?fd) → tomasys:fd_realisability(?fd, false)*
If a Component has a status in false (in error), and that Component is *required by* a Function Design then that *realisability* is set to *false*.

Table 1.
SWRL rules for proof-of-concept implementation. The first one sets the Function Grounding in error if it uses a faulty Component, the second one sets the Objective in error if the Function Grounding is in error and the third one marks as unreachable the Function Design that require unavailable Components.

DOI: http://dx.doi.org/10.5772/intechopen.97357

When the system needs adaptation because the mission (`Objective`) is in error according to rule no. 2, a selection of a design alternative is required. In this case, rule no. 3 is applied to determine the solution available in terms of components. An equivalent rule in terms of `Quality Attributes` and NFRs is used to select the design compliant with the mission requirements. If there are several `Function Designs` available, the module in charge of the reconfiguration, called Metacontroller, selects the `Function Design` with higher estimated performance.

In this case, each `Function Design` represent a system mode. For instance, the normal mode uses the laser to navigate at maximum velocity levels in environments with few obstacles but not crowded. The degraded mode, uses an RGBD camera instead of a laser to navigate, as the refresh rate of this device is considerably less than the laser, the velocity is reduced to keep navigation with safety. By contrast, the energy-saving mode is a very safe and slow implementation to reduce at maximum the battery consumption, impacting the duration of the mission, and therefore, the performance.

The ontology implementation has been evaluated in a complete robotic application, a patrolling corridors mission. While the robot performs the patrol, contingencies such as the laser error described previously. The robot used is a TurtleBot2 composed of a Kobuki platform RPLidar A2 laser and an Orbec Astra RGBD camera. The experiments consisted of simulating a laser error at a random instant. These data were corrupted by generating realistic data as if something was blocking the laser, or if there was a misalignment, maybe due to a hit or a fall of the robot. This was done by publishing scan messages with erroneous data (a vector of 0's) in the gazebo plugin topic. **Figure 2** depicts the simulation used to develop the reasoner to implement the ontology-driven reconfiguration. The output from the reasoner once the laser is malfunctioning and the robot with the navigation mode to degraded is shown in **Figure 3**.

The main experiment carried measures the recovery time for a laser failure using ontology-driven reconfiguration. After 50 iterations of the experiment, the time required is 1.995 s with a standard deviation of 0.478. Without it, the estimated recovery time for this failure is about 300 s (indeed tied to system maintenance).

Furthermore, another testbed has been used to prove the ontology reusability along with the reconfiguration performance. In this case, we have used a simulation of an unmanned underwater vehicle performing exploration in a flooded mine [27].

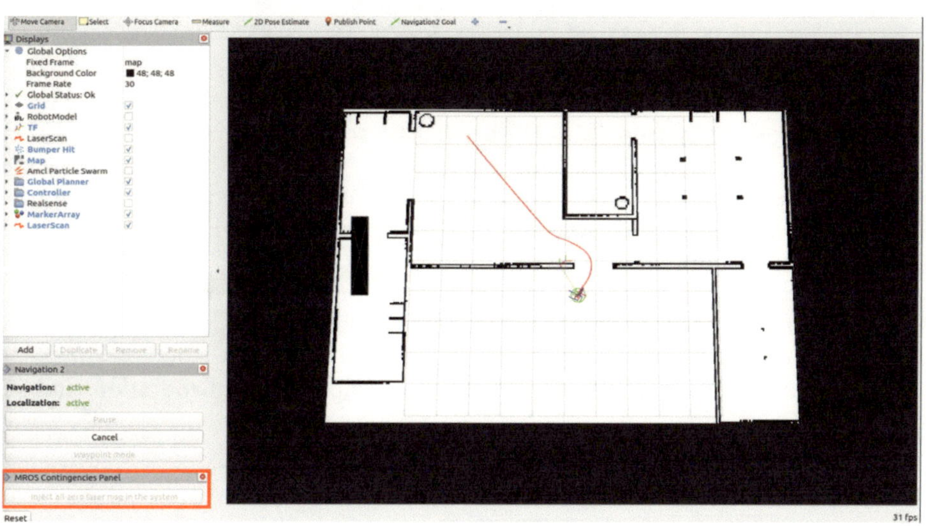

Figure 2.
Simulation of robot patrolling with navigation and localization system; in red, the button to inject laser failure.

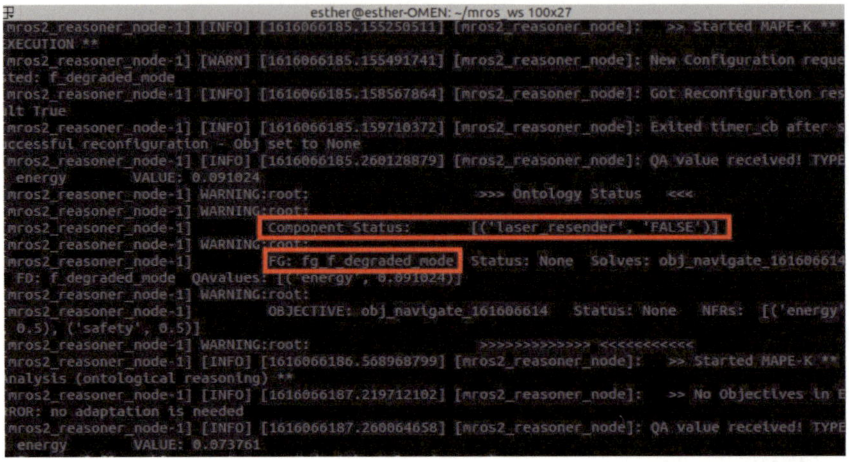

Figure 3.
Console from the reasoner ROS2 node to implement the ontology-driven reconfiguration; in red the component status of the laser as 'FALSE', malfunctioning and the navigation mode required, the Function Grounding 'DEGRADED MODE'.

With this proof-of-concept, we have implemented a fault-adaptive subsystem in general terms to ensure mission requirements and face component faults. This approach provides a general and reusable asset for systems in which there are design alternatives in terms of behavior and/or components. This approach aptly realizes the vision of using ontologies for building knowledge bases to decouple the mission-oriented system operation core from the reconfiguration needed to overcome disturbances or failures. Moreover, with this experimental setting, we have shown evidence of the advantages of automatic reconfiguration through ontological architectures for reducing the recovery time for laser contingencies.

6. Systemic implications of ontologies

As stated in Section 2, ontologies provide a variety of tools to define a system in terms of concepts and the relationships among them. Besides, given their formal nature, they can be included in the system as an explicit source of knowledge to improve its run-time operation.

Ontologies can be treated as a sub-system itself, that may be designed following the systems engineering principles of modularity, scalability, and reusability. The proof-of-concept presented here (Section 5), is an example of the use of ontologies to augment the autonomy level of a robot, increasing its dependability by improving mission assurance. In this proof-of-concept system ontologies are components of the deployed system.

However, the use of ontologies as part of autonomous systems engineering processes goes well beyond this [18], because they can have a strong impact on the many processes of the systems life cycle [28]. In this section, we analyze three classes of impacts: (i) impacts on complexity; (ii) impacts on collaboration; and (iii) impacts on risk management.

6.1 Implications on complexity

Section 2 summarised some of the benefits of modularity in ontologies. The key feature of the modular approach is the reduction obtained in the ontology complexity when working on small modules integrated into a whole ontology. This conceptual decomposition and complexity reduction can be extrapolated to the conceptualization that underlies all system engineering processes[7], especially in systems-of-systems contexts, [29].

The use of a knowledge base that makes explicit the requirements and the design decisions on the system, provides systems engineering knowledge that can be leveraged at the whole life cycle. The explicitness of mission-oriented concepts makes it easier for the developer at the verification phase to understand possible fault sources and integration problems. The trend in systems engineering towards model-based approaches is rooted in the formal verification capability that models provide at all stages (esp. at early stages where the costs of re-engineering are much lower). At the early stages of the model, complexity is lower and formal analyses may be more exhaustive and effective.

In the deployment phase—as shown in our proof-of-concept—ontologies may be used as fault-tolerance assets. The use of ontologies to reason about the state of the mission and the architectural components in use in general terms, provide a common framework to decouple reconfiguration actions from the particular implementation. This separation of concerns allows for a strong reduction in the complexity of the fault-tolerant mechanisms by both (i) the localized nature of the fault-tolerant mechanisms; and (ii) the possibility of reusing general tested assets such as TOMASys and some of the more general knowledge bases (as the ones presented here[8]).

6.2 Implications on collaborative systems

The use of ontologies in the construction of formal knowledge bases provides a common understanding within a domain. The encoding of ontologies in machine-readable formats such as OWL allows a truthful integration when sharing information between different agents and/or tools. This integration obviously includes the possibility of collaboration between different types of systems in a group mission. This collaboration is not limited to the activities in the systems life cycle as described earlier but spans all classes of multi-agent collaboration in fielded systems.

An example of this is the enabling of collaborative work between fleets of robots, especially when they are heterogeneous. For example, the shared information between an unmanned aerial vehicle (UAV) and an unmanned ground vehicle (UGV) may encounter incompatibilities just in the coordinate systems they use, as the UGV does not take into account the vertical axis. Moreover, robots may have different capabilities, a UAV may be able to map the environment whilst the UGV may have an arm to interact with it.

The same occurs when the system includes a human as an external operator or supervisor, the system must exchange information to ensure collaborative work. Usually, the combination of data is done implicitly by the system designers. However, using ontologies for formal integration of all the different perspectives of a system of systems with an explicit *conceptualization* provides a robust and trustworthy method for sharing information that affects the way humans interact and collaborate with machines.

6.3 Implications on risk management

As said before, a formal definition of system concepts can be used for better formal analysis along the whole life cycle. This upstream analysis implies a

---

[7] See for example http://www.sebok.org.

[8] Available at: https://github.com/MROS-RobMoSys-ITP/mros_ontology

reduction of the probability of faults during system operation but it also includes the possibility of better diagnostic features. having explicit knowledge on relationships among system entities allows the system to trace the source of faults and take action in the implicated parts.

Besides, the system may adapt itself as the case presented in Section 5 to reduce damage or can even take preventive action based on deep reasoning. When the system is able to adapt to overcome problems derived from environmental changes or malfunctioning components, the system becomes more autonomous and trustworthy.

Design decisions have always associated risks. Usually, the final design is commonly the solution with less risk in the context where the system is deployed. The use of ontologies provides a tool for risk management as design decisions may be justified in the knowledge base. Furthermore, this information can be used at run-time to apply the most suitable solutions if the operational environment changes.

In the proof-of-concept presented here, quality attributes are used to select among design alternatives. When risk augments because of not meeting the security standards of the mission, other designs can be used. This may affect the performance of the mission but ensures its fulfillment. The system engineer must coordinate adequately those quality attributes to ensure a trade-off between performance and security. In this context, risk ceases to be a collateral effect of system design and operation to become a first-level citizen in the explicit design of the system.

7. Towards a full life cycle ontology

Ontologies are commonly used to store information within a domain. For this reason, they are a valuable asset to model the shared understanding of the system and its concepts at different stages of its life cycle. Here, we propose to take a further step and use that information model at run-time to provide the system with knowledge-driven self-adaptation.

The proof-of-concept presented here addresses reconfiguration to ensure the mission fulfillment within a set of predefined operational requirements. The main limitation is imposed by the set of alternative designs predefined for the system and its possible contingencies. Here, ontologies have been used to orchestrate the deployment of different design solutions at run-time.

To increase the autonomy levels of systems, particularly in the case of robots, we propose to take a further step and include all early phases of system life cycles, from the need identification to fulfill a mission to the verification of the system (see **Figure 4**).

When the system is deployed, each time a contingency is detected and it requires reconfiguration, the system should return to the needs evaluation and analyze which tools it has available to satisfy the need. According to that, it can adopt some requirements and come across with a design by itself. That design may be tested in simulation with a digital twin or just deployed if it has been evaluated in previous operations.

To reach such an elevated autonomy level, a massive effort in ontology standardization in systems engineering is required[9]. Moreover, the design engineers must encode all the information they have about the system in terms of those

[9] See for example the efforts of the IEEE 1872.1 working group on robot task standardization or the IOF group on systems engineering.

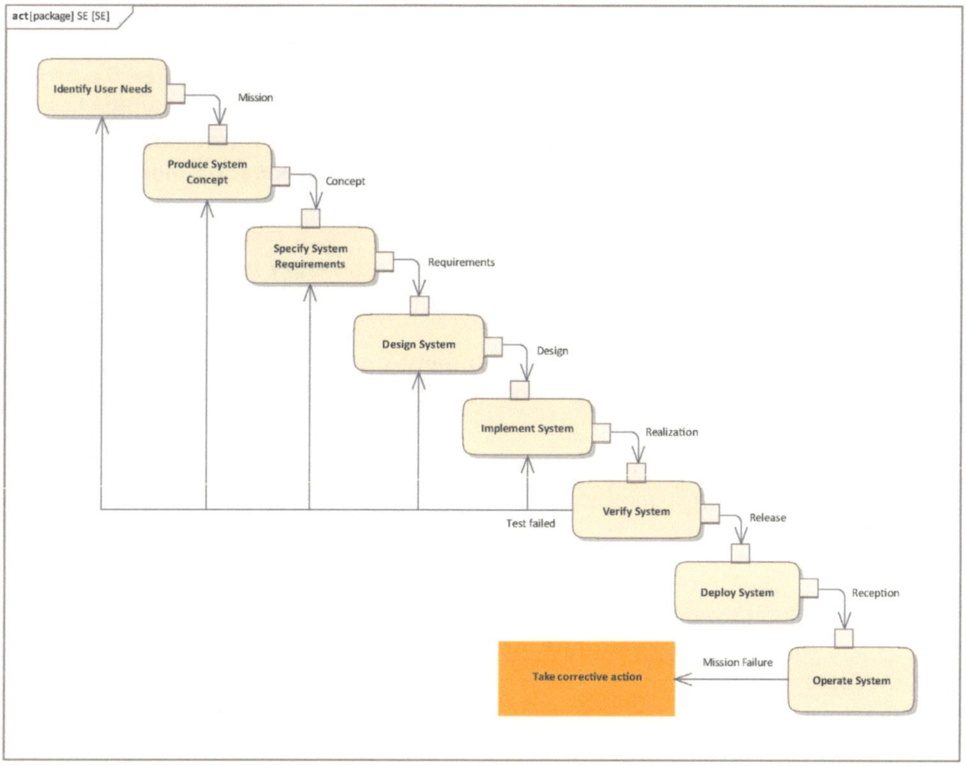

Figure 4.
Going back in the system life cycle from an operational failure.

standards. That information also includes discarded ideas and the reason why they are not included in the final design, as can become part of a contingency solution.

To select the best possible design available in case of contingency, knowledge bases should include metrics regarding the cost (in terms of time, energy consumption, reliability) of the resulting system after applying reconfiguration. This selection must preserve the mission fulfillment with optimal features.

8. Conclusions

Ontologies provide a baseline for shared information between systems, subsystems, and external agents. But that information can also be a tool for augmenting the autonomy levels of systems. Ontologies provide a formal conceptualization of entities and their relationships. The knowledge stored can be used along architectural models in system engineering to provide a general framework for autonomous systems.

A concrete realization of this general framework has been tested with a mobile robot navigating to a point in a cluttered environment. The proof-of-concept address two contingency types, (i) a component-level failure, as the case when the laser becomes unavailable, or (ii) not reaching the mission requirements in terms of safety or energy consumption.

The contingencies are solved by reasoning about the status of the components in use and the status of the objective and its quality attributes associated. Once a contingency is encountered, the reasoner provides the most suitable design alternative to overcome it according to the component availability and the estimated quality values in terms of energy consumption, safety and performance.

However, this proof-of-concept focus on the selection of alternatives. To provide a complete module of self-adaptation and reconfiguration in the whole life cycle of systems, we need to take a larger perspective. To reach complete autonomy, the system needs to have access to knowledge from the needs for which it is designed, besides the design alternatives and the restrictions the engineers have faced. With this information, in case of failure, the system can have a wider picture of its context and take corrective action at different levels of its architecture to adapt to run-time situations.

To achieve this vision a *full system life cycle ontology for autonomous systems* is needed. Current efforts in ontology development for systems engineering, robots, and autonomous systems are quite valuable but they shall be (i) based on a general systems foundation; (ii) harmonized, and (iii) built with a modular ontology approach.

Acknowledgements

This work was supported by RobMoSys-ITP-MROS (Grant Agreement No. 732410) and ROBOMINERS (Grant Agreement No. 820971) projects with funding from the European Union's Horizon 2020 research and innovation programme.

Author details

Esther Aguado* and Ricardo Sanz
Autonomous Systems Laboratory, Centre for Automation and Robotics UPM-CSIC, Universidad Politécnica de Madrid, Spain

*Address all correspondence to: e.aguado@upm.es

IntechOpen

© 2021 The Author(s). Licensee IntechOpen. This chapter is distributed under the terms of the Creative Commons Attribution License (http://creativecommons.org/licenses/by/3.0), which permits unrestricted use, distribution, and reproduction in any medium, provided the original work is properly cited. (cc) BY

References

[1] Gruber TR. A translation approach to portable ontologies. Knowledge Acquisition. 1993;5(2):199–220.

[2] Guarino N, Giaretta P. Ontologies and knowledge bases: Towards a terminological clarification. In: Towards very Large Knowledge bases: Knowledge Building and Knowledge sharing. IOS Press; 1995. p. 25–32.

[3] Zhao X, Wang Z, Cui Y, Zheng G. Novel Ontology-Based Method for Generating Satellite Cluster's Task Configuration. Journal of Aerospace Information Systems. 2020;17(2):86–96.

[4] Poggi F, Rossi D, Ciancarini P. Integrating Semantic Run-Time Models for Adaptive Software Systems. In: Journal of Web Engineering. vol. 18; 2019. p. 1–42.

[5] Zhai Z, Martínez Ortega JF, Lucas Martínez N, Castillejo P. A Rule-Based Reasoner for Underwater Robots Using OWL and SWRL. Sensors (Basel, Switzerland). 2018 10;18(10):3481.

[6] IEEE Standard Ontologies for Robotics and Automation. IEEE Std 1872-2015. 2015:1–60.

[7] Prestes E, Carbonera JL, Rama Fiorini S, M Jorge VA, Abel M, Madhavan R, et al. Towards a core ontology for robotics and automation. Robotics and Autonomous Systems. 2013; 61(11):1193 – 1204. Ubiquitous Robotics.

[8] Niles I, Pease A. Towards a Standard Upper Ontology. In: Proceedings of the International Conference on Formal Ontology in Information Systems - Volume 2001. FOIS '01. New York, NY, USA: Association for Computing Machinery; 2001. p. 2–9.

[9] Carbonera JL, Fiorini SR, Prestes E, Jorge VAM, Abel M, Madhavan R, et al. Defining positioning in a core ontology for robotics. In: 2013 IEEE/RSJ International Conference on Intelligent Robots and Systems; 2013. p. 1867–1872.

[10] Olszewska JI, Barreto M, Bermejo-Alonso J, Carbonera J, Chibani A, Fiorini S, et al. Ontology for autonomous robotics. In: 2017 26th IEEE International Symposium on Robot and Human Interactive Communication (RO-MAN); 2017. p. 189–194.

[11] Fiorini SR, Carbonera JL, Gonc¸alves P, Jorge VAM, Rey VF, Haidegger T, et al. Extensions to the core ontology for robotics and automation. Robotics and Computer-Integrated Manufacturing. 2015;33: 3 – 11. Special Issue on Knowledge Driven Robotics and Manufacturing.

[12] Parent C, Spaccapietra S. In: Stuckenschmidt H, Parent C, Spaccapietra S, editors. An Overview of Modularity. Berlin, Heidelberg: Springer Berlin Heidelberg; 2009. p. 5–23.

[13] Olivares-Alarcos A, Beßler D, Khamis A, Goncalves P, Habib MK, Bermejo J, et al. A review and comparison of ontology-based approaches to robot autonomy. The Knowledge Engineering Review. 2019;34.

[14] Ragavan SV, Ganapathy V. A General Telematics Framework for Autonomous Service Robots. In: 2007 IEEE International Conference on Automation Science and Engineering; 2007. p. 609–614.

[15] Bayat B, Bermejo J, Carbonera J, Facchinetti T, Fiorini S, Gonçalves P, et al. Requirements for building an ontology for autonomous robots. Industrial Robot: An International Journal. 2016 08;43.

[16] Kleppe AG, Warmer J, Bast W. MDA Explained: The Model Driven Architecture: Practice and Promise.

USA: Addison-Wesley Longman Publishing Co., Inc.; 2003.

[17] Hernández C, Bermejo-Alonso J, Sanz R. A self-adaptation framework based on functional knowledge for augmented autonomy in robots. Integrated Computer-Aided Engineering. 2018;25:157–172.

[18] Sanz R, Bermejo J, Morago J, Hern´ andez C. Ontologies as Backbone of Cognitive Systems Engineering. In: Bryson J, Vos MD, Padget J, editors. Proceedings of AISB CAOS 2017: Cognition And Ontologies. Bath, UK; 2017. p. 218–223.

[19] Sanz R, Matía F, Galán S. Fridges, Elephants and the Meaning of Autonomy and Intelligence. In: Groumpos PP, Koussoulas NT, Polycarpou M, editors. IEEE International Symposium on Intelligent Control, ISIC'2000. Patras, Greece; 2000. p. 217 – 222.

[20] López I, Sanz R, Hernández C, Hernando A. General Autonomous Systems: The Principle of Minimal Structure. In: Grzech A, editor. Proceedings of the 16th International Conference on Systems Science. vol. 1; 2007. p. 198–203.

[21] López I. A Framework for Perception in Autonomous Systems [Ph.D. thesis]. Departamento de Automática, ETS de Ingenieros Industriales, Universidad Politécnica de Madrid; 2007.

[22] Klir GC. An Approach to General Systems Theory. Van Nostrand Reinhold; 1969.

[23] Abbass HA, Scholz J, Reid DJ. In: Abbass HA, Scholz J, Reid DJ, editors. Foundations of Trusted Autonomy: An Introduction. Cham: Springer International Publishing; 2018. p. 1–12.

[24] Amaral G, Guizzardi G, Guizzardi R, Mylopoulos J. Ontology-based Modeling and Analysis of Trustworthiness Requirements: Preliminary Results. In: Dobbie G, Frank U, Kappel G, Liddle SW, Mayr HC, editors. International Conference on Conceptual Modeling (ER 2020). Vienna, Austria; 2020. p. 342–352.

[25] Kok B, Soh H. Trust in Robots: Challenges and Opportunities. Current Robotics Reports. 2020 12;1:1–13.

[26] W3C. OWL 2 Web Ontology Language Structural Specification and Functional-Style Syntax (Second Edition). World Wide Web Consortium; 2012. REC-owl2-syntax-20121211.

[27] Aguado E, Milosevic Z, Hernández C, Sanz R, Garzon M, Bozhinoski D, et al. Functional Self-Awareness and Metacontrol for Underwater Robot Autonomy. Sensors. 2021;21(4). Available from: https://www.mdpi.com/1424-8220/21/4/1210.

[28] ISO/IEC/IEEE. ISO/IEC/IEEE 15288-2015 Systems and software engineering – System life cycle processes. International Standards Organisation; 2015.

[29] Guariniello C, Raz AK, Fang Z, DeLaurentis D. System-of-systems tools and techniques for the analysis of cyber-physical systems. Systems Engineering. 2020;23(4):480–491.

Chapter 6

System Level Design and Conception of a System-on-a-Chip (SoC) for Cognitive Robotics

Diego Stéfano Fonseca Ferreira, Augusto Loureiro da Costa, Wagner Luiz Alves De Oliveira and Alejandro Rafael Garcia Ramirez

Abstract

In this work, a system level design and conception of a System-on-a-Chip (SoC) for the execution of cognitive agents in robotics will be presented. The cognitive model of the Concurrent Autonomous Agent (CAA), which was already successfully applied in several robotics applications, is used as a reference for the development of the hardware architecture. This cognitive model comprises three levels that run concurrently, namely the reactive level (perception-action cycle that executes predefined behaviours), the instinctive level (receives goals from cognitive level and uses a knowledge based system for selecting behaviours in the reactive level) and the cognitive level (planning). For the development of such system level hardware model, the C++ library SystemC with Transaction Level Modelling (TLM) 2.0 will be used. A system model of a module that executes a knowledge based system is presented, followed by a system level description of a processor dedicated to the execution of the *Graphplan* planning algorithm. The buses interconnecting these modules are modelled by the TLM generic payload. Results from simulated experiments with complex knowledge bases for solving planning problems in different robotics contexts demonstrate the correctness of the proposed architecture. Finally, a discussion on performance gains takes place in the end.

Keywords: Autonomous Agents, Robotics, Hardware Design, Knowledge Based Systems, Transaction Level Modelling

1. Introduction

Behaviour-based robotics is a branch of robotics that studies techniques for the interaction of robotic agents with the environment using the perception-action cycle in a coordinated fashion. With the addition of cognition, these agents may use knowledge about the environment to perform more complex tasks [1–3]. In the context of artificial intelligence, the internal structure of those agents, i. e., their cognitive architectures, dictate how the problem-solving will take place [4].

An example of a cognitive architecture with successful applications in robotics is the Concurrent Autonomous Agent (CAA), an autonomous agent architecture for

IntechOpen

mobile robots that has already proven to be very powerful [5–7]. This agent possesses a three layer architecture, in which each layer is responsible for a different task: the reactive layer runs behaviours with a perception-action cycle; the instinctive layer coordinates reactive behaviour selection; and the cognitive layer does the high-level planning.

In this work, a system level hardware model of a System-on-a-Chip (SoC) for cognitive agents will be presented. This model was inspired by the cognitive architecture of the CAA. Therefore, the CAA will be described in Section 2. In Sections 3 and 4 the *Rete* and *Graphplan* algorithms are described, respectively, since they are at the core of the CAA. The SystemC and TLM 2.0 standards, the tools used to construct the models are presented in Section 5, followed by the presentation of the proposed architecture in Section 6. Results of experiments are shown in Section 7 and some final thoughts and conclusions are presented in Section 8.

2. The concurrent autonomous agent (CAA)

The Concurrent Autonomous Agent (CAA) is a cognitive architecture whose taxonomy is based on the generic model for cognitive agents, which is composed by the reactive, the instinctive and the cognitive levels [8]. The CAA levels are presented in **Figure 1** [5, 6], where the message passing between the levels is shown. The cognitive level generates plans that are executed by the instinctive level through the selection of reactive behaviours in the reactive level. [6].

Both the cognitive and instinctive levels apply a Knowledge Based System (KBS) for knowledge representation and inference. The KBS is composed by a facts base, a rules base and an inference engine, as shown in **Figure 2** [6].

The facts base contains atomic logical elements that represents knowledge that is known about the current state of the environment. The rules base contains a set of rules in the format `if PREMISE then CONSEQUENT`. The premise consists of a conjunction of ungrounded fact patterns that uses variables to increase expressiveness.

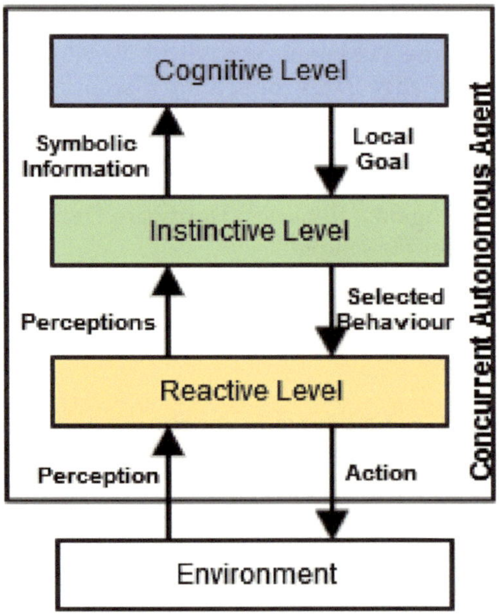

Figure 1.
Cocurrent autonomous agent architecture.

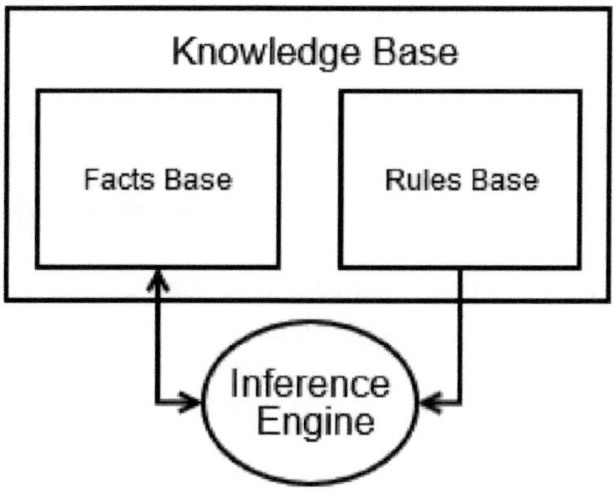

Figure 2.
KBS used by AAC.

The consequent, in turn, has instructions on how to modify the facts base and which message should be sent to other levels, if any. The KBS then goes through the following cycle [9].

- Recognition: identify which rules can be activated by checking if the premises matches the facts in the facts base;
- Conflict Resolution: among the activated rules (conflict resolution set), decide which should be executed; and
- Execution: the chosen rule in the conflict resolution phase has its consequent executed.

The *Rete* matching algorithm is applied in the recognition phase to generate the conflict resolution set. The instinctive level uses its KBS to select the appropriate reactive behaviour to be selected given the current world state. The cognitive level, in turn, uses its KBS inside the *Graphplan* algorithm (that will be described later in this chapter), in the state space expansion stage [10].

3. The *Rete* algorithm

As mentioned earlier in this chapter, the *Rete* matching algorithm is employed in the recognition stage the KBSs used by the CAA. It is proposed in [11], and is named after the latin word for "network".

The algorithm builds a graph out of the rules base of the KBS where each node has a special purpose. In the end, it avoids running through the entire facts base for each rule premise, every time a new fact arrives, by saving information about partial matches in some of its nodes [11].

The constructed graph has two portions: the alpha network, responsible for comparing the constants in the premises with the corresponding fields in the incoming fact; and the beta network, which checks for variable assignment consistency and maintenance of partial matches [10].

The nodes the compose the alpha network are the following [10]:

- *Root Node*: entry point for new facts;

- *Constant Test Nodes* (CTN): compares constant fields of premises with corresponding ones in the current fact; and
- *Alpha Memories* (AM): stores facts that successfully passed the tests in CTNs.

The beta network is composed by the following nodes [10]:

- *Join Nodes* (JN): perform tests that ensure variable assignment consistency inside a premise instance (partial match);
- *Beta Memories* (BM): stores partial matches produced in JNs; and
- *Production Nodes*: terminal nodes for full rule matches.

4. The *Graphplan* algorithm

The cognitive level uses the *Graphplan* algorithm to generate the plans that the other levels should execute. Originally, the algorithm used a propositional knowledge representation, so this will be adopted here for the algorithm description. The rest of this section uses [12, 13] as references.

Mathematically, a planning problem may be stated as $\mathcal{P} = (\Sigma, s_j, g)$, where $\Sigma = (S, A, \gamma)$ is the problem domain (that comprises the set of states S, the set of actions A and a state transformation function $\gamma = S \times A \rightarrow S$), s_j is the initial state and g is the goal state.

Each action $a \in A$ has a set $prencond(a)$ of precondition propositions and a set $effects(a) = effects^+(a) \cup effects^-(a)$ of effects. The effects, in turn, may be broken down into two subsets: $effects^+(a)$, the set of positive propositions (propositions to be added), and $effects^-(a)$, the set of negative propositions (propositions to be deleted). The applicability condition for an action a, in a given state s, may be written as $precond(a) \subseteq s$. The new state produced by the application of a would be $\gamma(s, a) = (s - effects^-(a)) \cup effects^+(a)$.

Consider an action layer A_j and the propositional layer P_{j-1} preceding it. A_j contains all actions a such that $precond(a) \subseteq P_{j-1}$, and P_{j-1} all propositions p such that $p \in P_{j-1}$. The so called planning graph is the built by connecting elements in P_{j-1} to elements in A_j by edges:

- edges connecting a proposition $p \in P_{j-1}$ to an action $a \in A_j$;
- edges connecting an action $a \in A_j$ to a proposition $p \in P_{j-1}$, such that $p \in effects^+(a)$ (positive arc); and
- edges connecting an action $a \in A_j$ to a proposition $p \in P_{j-1}$, such that $p \in effects^-(a)$ (negative arc).

If two actions $a_1, a_2 \in A_j$ obey $effects^-(a_1) \cap (precond(a_2) \cup effects^+(a_2)) = \varnothing$ and $effects^-(a_2) \cap (precond(a_1) \cup effects^+(a_1)) = \varnothing$, they a said to be *independent*; if not, they are *dependent*, or *mutually exclusive* (*mutex*).

Propositions can also be *mutex*: p and q are *mutex* if every action in A_j that adds p is *mutex* with every action in A_j that produces q, and there are no actions in A_j that adds both p and q. Also, if a precondition of an action is *mutex* with a precondition of another action, the actions are *mutex*.

The algorithm begins by expanding the graph. The pseudo-code for the expansion step is given in Algorithm 1.

Algorithm 1 Planning graph expansion

procedure EXPAND (s_i) $\quad\triangleright s_i$: i-th state layer
  $A_{i+1} \leftarrow KBS.InferenceCycle(s_i, A)$ $\quad\triangleright A$: action profiles
  $s_{i+1} \leftarrow \cup A_{i+1}.effects^{+}$
  $\mu A_{i+1} \leftarrow \{(a,b) \in A^2_{i+1}, a \neq b \mid Dependent(a,b) \vee \exists (p,q) \in \mu s_i : p \in preconds(a), q \in preconds(b)\}$
  $\mu s_{i+1} \leftarrow \{(p,q) \in s^2_{i+1}, p \neq q \mid \forall (a,b) \in A^2_{i+1} : p \in effects^{+}(a) \wedge q \in effects^{+}(b) \rightarrow (a,b) \in \mu A_{i+1}\}$
end procedure

The expansion stops when the goal state g is detected in the state layer s_i. It then triggers a recursive search for non-*mutex* actions in all the preceding action layers that could have produced the goal state found in s_i. This procedure is composed by the functions Search (Algorithm 2) and Extract (Algorithm 3).

Algorithm 2 Search for *non-mutex* actions.

procedure SEARCH(g, π_i, i)
  if $g = \varnothing$ **then**
    $\Pi \leftarrow Extract(\cup\{preconds(a) \mid \forall a \in \pi_i\}, i-1)$
    if $\Pi = Failure$ **then**
      return *Failure*
    end if
    return $\Pi.\pi_i$
  else
    select any $p \in g$
    $resolvers \leftarrow \{a \in A_i \mid p \in effects^{+}(a) \wedge \forall b \in \pi_i : (a,b) \notin \mu A_i\}$
    if $resolvers = \varnothing$ **then**
      return *Failure*
    end if
    nondeterministically choose $a \in resolvers$
    return Search$(g - effects^{+}(a), \pi_i \cup \{a\}, i)$
  end if
end procedure

Algorithm 3 Extract a plan.

procedure EXTRACT(g, i)
  if $i = 0$ **then**
    return $\varnothing$
  end if
  $\pi_i \leftarrow Search(g, \varnothing, i)$
  if $\pi_i \neq Failure$ **then**
    return π_i
  end if
  return *Failure*
end procedure

5. SystemC and transaction level modelling

This work uses SystemC and the TLM 2.0 as modelling and simulation tools, so in this section they will be described.

5.1 SystemC

In the design of complex digital systems, obtaining a high-level executable specification of the project in early stages of the design process is useful for detecting errors or validate functionality prior to implementation. This is one of the main advantages of SystemC, a C++ class library for hardware design at various abstraction levels - from system level to Register Transfer Level (RTL). **Figure 3** shows the typical design flow for SystemC projects [14].

The SystemC library contains elements that facilitates representation of hardware systems parallelism. Hardware models in SystemC are represented by modules that may run in parallel interconnected by ports and channels (**Figure 4**). In this way, the initial model may contain a few modules representing system level

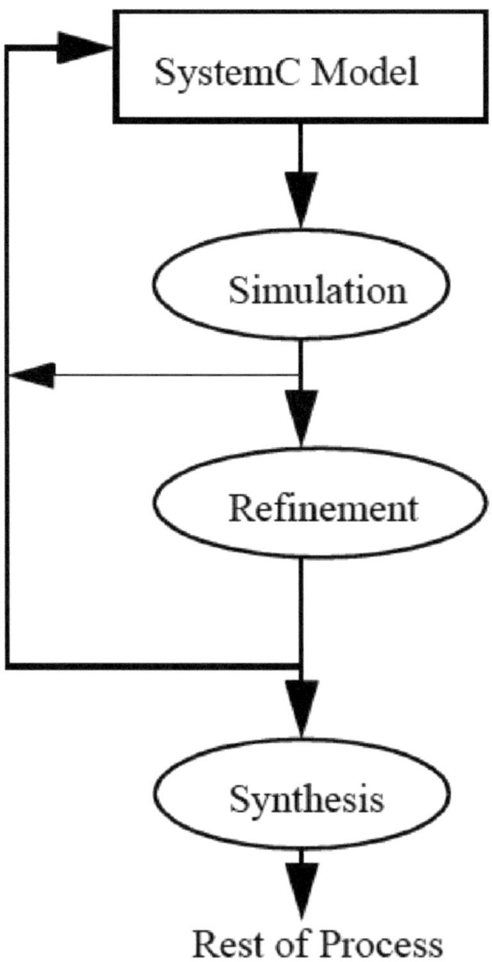

Figure 3.
SystemC hardware design flow [14].

functionality and, as the model gets refined, those initial high-level modules are further divided into more specific interconnected modules, until the RTL is reached [14].

5.2 Transaction level modelling

In hardware models of higher levels of abstraction, executing all modules at each time step may produce an unnecessary overhead. Thinking of a digital systems as components connected by a bus, reading from and writing to it, it would be more efficient to execute modules only when they have something massages to send/ receive. This is the rationale behind Transaction Level Modelling (TLM), the message exchange being called a transaction [15].

With SystemC, an implementation of the TLM called TLM 2.0 is provided. It inherits all the SystemC capabilities, mainly the module concept, extending it with sockets, transactions and payloads (**Figure 5**).

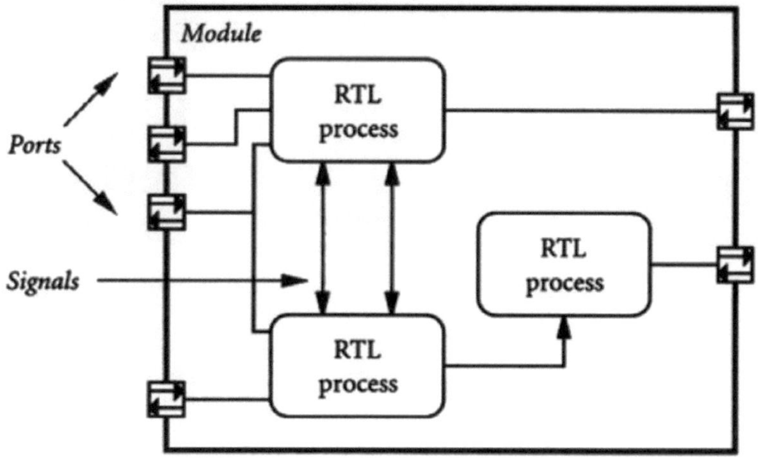

Figure 4.
Typical SystemC RTL module [14].

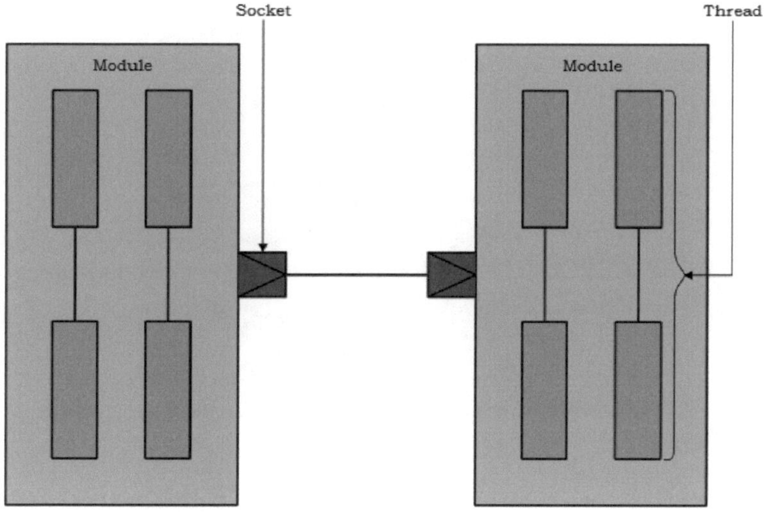

Figure 5.
TLM basic elements [15].

6. Proposed architecture

The hardware architecture proposed in this work takes full advantage of SystemC and TLM 2.0 capability of developing executable specifications from system level to RTL. In this sense, the approach employed was to obtain a high level model and validate its functionality using experiments in a robotics context.

The TLM model proposed is shown in **Figure 6**. It consists of modified SystemC model of the *Rete* processor presented in the authors previous work [10]. As can be seen in **Figure 6**, the instinctive module now implements a detailed *Rete* processor, that uses two Content Addressable Memories (CAMs) to implement the knowledge base and an auxiliary stage for test execution.

The Instruction Set Architecture (ISA) of the *Rete* processor described in [10] is still employed in this model, but now some tasks related to join node in the Rete network are performed separately in the Join Node Module.

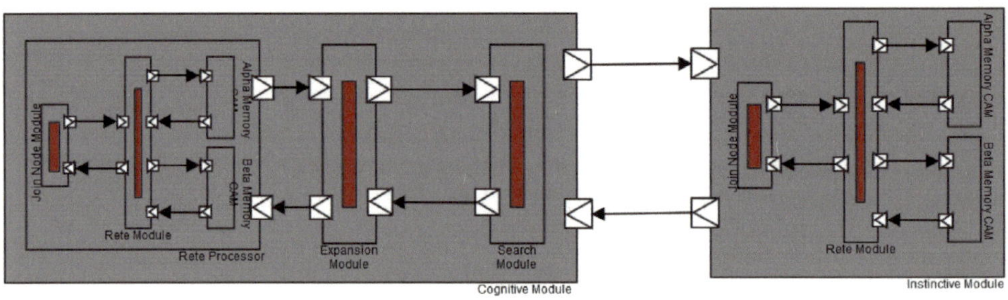

Figure 6.
TLM model of the SoC.

7. Results

7.1 Problem domain and simulation environment

The experiments were performed using the Webots R2021a robotics simulator. In the context of the CAA, the reactive level of the agent was

Figure 7.
Simulation environment for start state.

implemented inside this simulator, in the form of behaviours and controllers the interface with the environment. In the simulator, the planning problem domain was constructed: a simplified version of the blocks domain. The simulation consisted of three coloured boxes (red, green and blue) disposed in a given order around KUKA Youbot robot, which is a mobile robot with a robotic arm and a plate. The simulation environment and the robot in the initial state are shown in the **Figure 7**.

The planning problem consisted of reordering the blocks from the initial position shown in **Figure 7** so that the red block is in the left side or the arm, the blue in the right and the green in the front.

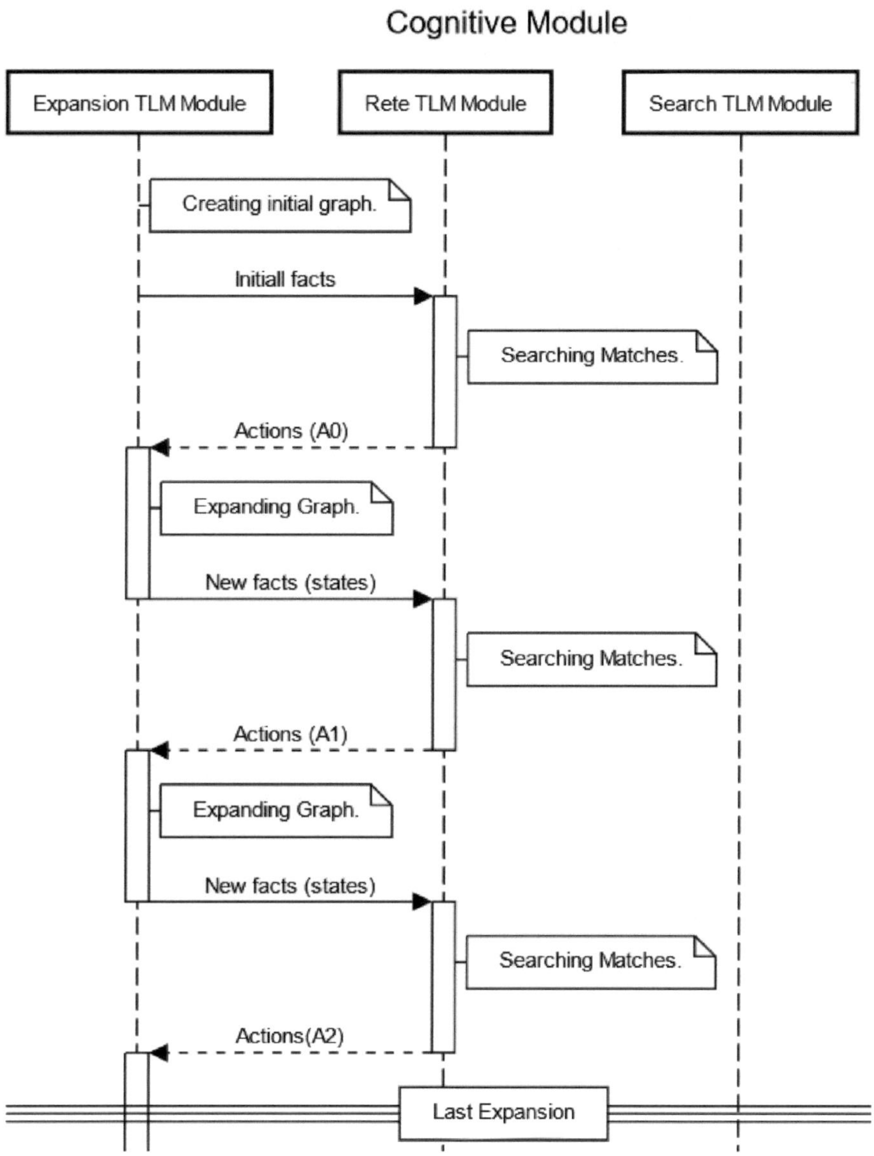

Figure 8.
Sequence diagram for graph expansion.

7.2 Cognitive module results

The operation of this module will be presented through a sequence diagram. The first part of this diagram, shown in **Figure 8**, shows how this level expanded the planning graph until what was labelled as last expansion.

The *Rete* and Expansion TLM modules together expand the planning graph: the current state is given as an input for the *Rete* module, that gives in return the next action layer. This is done 3 times, until action layer A2 is reached. The Expansion Module then processes the consequence of the newly added actions, updating the state layer. But this time, the goal state is present in the state layer, so a transaction is sent to the search module to backtrack the goal state checking if the actions that produced are mutex with any other. If no mutex relation is found, those actions form the plan. And, as shown in **Figure 9**, this plan is, indeed, found in the first backtrack attempt.

As can be seen in **Figure 9**, during the search for a solution, the expansion continues to take place, but is interrupted when the Search Module reports the solution. The plan found to the given problem was composed by the actions *move (green, right, front)*, *move(blue, left, right)* and *move(red, back, left)*.

7.3 Instinctive module results

The reactive behaviours for the robotic arm were defined as: going to a reset position; moving to left, right, front or back; grip and release. In order to execute the actions produced by the cognitive level, a knowledge base was created and

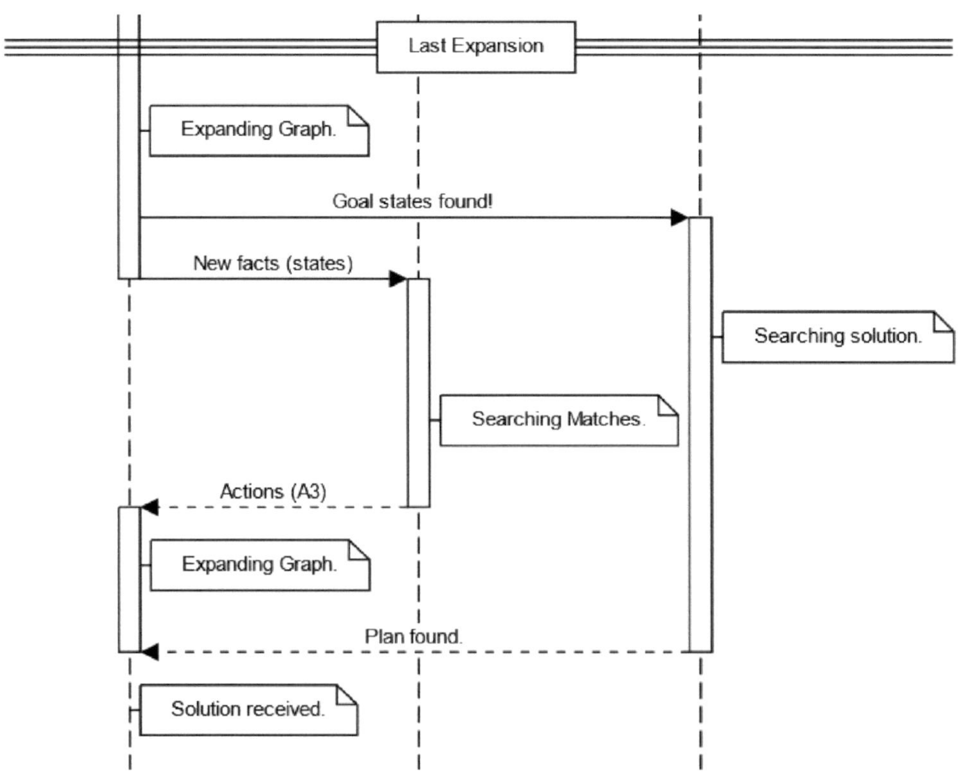

Figure 9.
*Sequence diagram for finding a plan (continuation of **Figure 8**).*

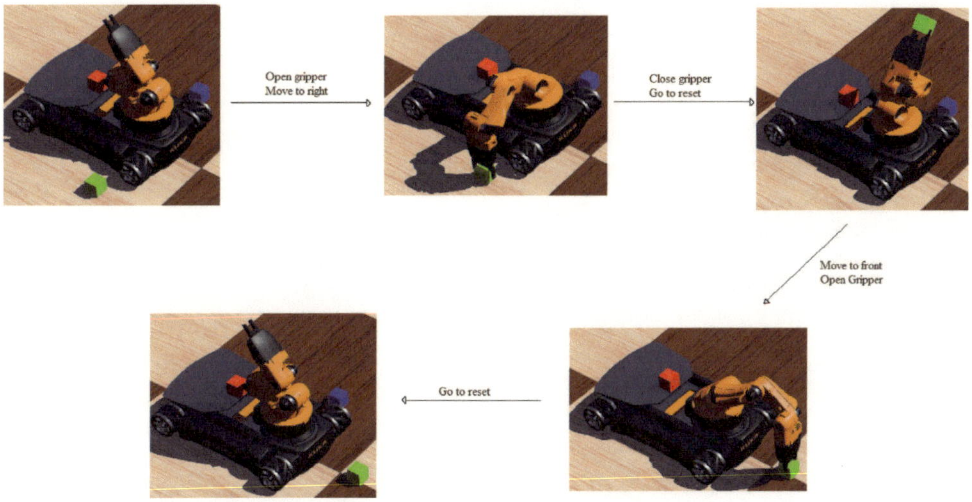

Figure 10.
Sequence of arm configurations and the reactive behaviours executed between them.

compiled for the *Rete* processor using its application specific ISA. The rules in this knowledge based were "grab" and "put". Both has as precondition that the arm is in the reset position and variables to specify the side where to grab from and the side where to put. The sequence of reactive behaviours activate by the instinctive level is show for the execution of the first action of the plan (*move(green, right, front)*) in **Figure 10**.

8. Conclusions

This chapter presented a SoC for cognitive agents that can perform symbolic computations at the hardware level. The cognitive model of the CAA was used as a reference for the hardware system-level model development, mapping its instinctive level to module with an application specific processor that executes the *Rete* matching algorithm, and with its cognitive level mapped into a module specifically designed for running the *Graphplan* planning algorithm (also with the use of the *Rete* processor). The SystemC and the TLM were used to build executable specifications that could validate its functionality in a robotics context. This version of the model was presented in a unified fashion, using SystemC/TLM modules and threads for the executable specification generation.

The results shown that the planning problem was solved by the Cognitive Module of the proposed architecture and successfully executed by its Instinctive Module, that consists of a *Rete* processor. By using a parallel architecture, the Cognitive Module broke the planning task into concurrent tasks in such a way that the backtrack search of the plan could take place while the graph were still expanding, as shown in **Figure 9**. In a complex planning problem this is advantageous because the solution usually does not come from the first backtrack search; thus, by not stalling the graph expansion, performance is gained.

In future works, tests with more complex knowledge bases and planning domains will be performed. Also, further refinements should be made in the architecture aiming synthesis.

Author details

Diego Stéfano Fonseca Ferreira[1†], Augusto Loureiro da Costa[1*†], Wagner Luiz Alves De Oliveira[1†] and Alejandro Rafael Garcia Ramirez[2†]

1 Robotics Laboratory, Electrical Engineering Department, Federal University of Bahia, Salvador, BA, Brazil

2 Computer Engineering Department, University of Vale de Itajaí, Itajaí, SC, Brazil

*Address all correspondence to: augusto.loureiro@ufba.br

† These authors contributed equally.

IntechOpen

© 2021 The Author(s). Licensee IntechOpen. This chapter is distributed under the terms of the Creative Commons Attribution License (http://creativecommons.org/licenses/by/3.0), which permits unrestricted use, distribution, and reproduction in any medium, provided the original work is properly cited.

References

[1] Huhns MN, Singh MP. Cognitive agents. Internet Computing, IEEE. 1998; 2(6):87–89.

[2] Guzel MS, Bicker R. A behaviour-based architecture for mapless navigation using vision. International Journal of Advanced Robotic Systems. 2012;9(1):18.

[3] Nattharith P, Güzel MS. Machine vision and fuzzy logic-based navigation control of a goal-oriented mobile robot. Adaptive Behavior. 2016;24(3):168–180.

[4] Langley P, Laird JE, Rogers S. Cognitive architectures: Research issues and challenges. Cognitive Systems Research. 2009;10(2):141–160.

[5] Costa ALd, Bittencourt G. From a concurrent architecture to a concurrent autonomous agents architecture. Lecture Notes in Artificial Inteligence. 1999;1856:85–90.

[6] Bittencourt G, Costa ALd. Hybrid Cognitive Model. In: The Third International Conference on Cognitive Science ICCS'2001 Workshop on Cognitive Angents and Agent Interaction; 2001.

[7] Cerqueira RG, Costa ALd, McGill SG, Lee D, Pappas G. From reactive to cognitive agents: Extending reinforcement learning to generate symbolic knowledge bases. In: Simpósio Brasileiro de Automação Inteligente 2013; 2013.

[8] Bittencourt G. In the quest of the missing link. International Joint Conference of Artificial Intelligence. 1997.

[9] Brachman R, Levesque H. Knowledge Representation and Reasoning. Elsevier; 2004.

[10] Ferreira D, da Costa AL, De Oliveira WLA. IntelliSoC: A system level design and conception of a system-on-a-Chip (SoC) to cognitive agents architecture. Applications of Mobile Robots. 2019:199.

[11] Forgy CL. On the Efficient Implementation of Production Systems. Carnegie-Mellon University; 1979.

[12] Ghallab M, Nau D, Traverso P. Automated Planning: Theory and Practice. Elsevier; 2004.

[13] Blum AL, Furst ML. Fast planning through planning graph analysis. Artificial intelligence. 1997;90(1):281–300.

[14] Synopsys I. SystemC 2.0 User's Guide; 2002].

[15] Bennett J. Building a Loosely Timed SoC Model with OSCI TLM 2.0. embecosm Application Note 1. 2008;(1).

Chapter 7

Quadrotor Unmanned Aerial Vehicles: Visual Interface for Simulation and Control Development

Manuel A. Rendón

Abstract

Quadrotor control is an exciting research area. Despite last years developments, some aspects demand a deeper analysis: How a quadrotor operates in challenging trajectories, how to define trajectory limits, or how changing physical characteristics of the device affects the performance. A visual interface development platform is a valuable tool to support this effort, and one of these tools is briefly described in this Chapter. The quadrotor model uses Newton-Euler equations with Euler angles, and considers the effect of air drag and propellers' speed dynamics, as well as measurement noise and limits for propeller speeds. The tool is able to test any device just by setting a few parameters. A three-dimensional optimal trajectory defined by a set of waypoints and corresponding times, is calculated with the help of a *Minimum Snap Trajectory* planning algorithm. Small Angle Control, Desired Thrust Vector (DTV) Control and Geometric Tracking Control are the available strategies in the tool for quadrotor attitude and trajectory following control. The control gains are calculated using *Particle Swarm Optimization*. Root Mean Square (RMS) error and Basin of Attraction are employed for validation. The tool allows to choose the control strategy by visual evaluation on a graphical user interface (GUI), or analyzing the numerical results. The tool is modular and open to other control strategies, and is available in GitHub.

Keywords: Quadrotor, Trajectory Planning, Trajectory Tracking

1. Introduction

Quadrotors are a special type of unmanned aerial vehicles (UAVs), increasingly employed last years for mapping, surveillance, searching and tracking operations, in rescue missions, agriculture, traffic management, landscape film making, and others [1–3].

When quadrotors applications demand large angle attitude control and obstacle avoidance, the following areas still need to be strengthened: Aggressive maneuvering control, visual-based control, localization in indoor environments, optimizing the computational cost for complex algorithms, and fault-tolerant disturbance rejection [4]. Several controllers have been proposed for these tasks: classic techniques as PD [5] and PID [6], optimal techniques as LQR [7] and LQG [8], nonlinear techniques such as Lyapunov [9] and *Backstepping* [5, 9] and intelligent and adaptive control techniques as Fuzzy200 [10] and Reinforcement Learning [11].

Research in quadrotor demand sophisticated equipment and costly laboratory. However, it can be optimized employing low cost virtual development tools. In autonomous control the path planning, path tracking and joint operation with other UAVs, can be supported by optimization techniques with support of software tools [1, 2, 12]. Published works relate the use of software such as Visual Basic, MatLab, Panda3D, Gazebo, or more open applications developed in languages as Python and C+ + [2]. In [1], a UAV 3D flight environment programmed in Python and developed in Panda3D is presented. A GUI developed in LabVIEW was published in [2], and other developed on MatLab-Simulink employs quadratic linear regulator (LQR) control [13].

Gazebo is an important virtual environment for robotics. Its integration with ROS provides a powerful testbed to analyze control algorithms. In [12] works that employ Gazebo and ROS for developing simulation of UAVs are described.

Most of the cited simulation tools have a difficult start for users with little programming experience. Even open-source models can be tricky. An interesting tool depicted in [14, 15] gives support for quadrotor control development, analyzing and comparing various control strategies on challenging trajectories. It may be used by beginning users with not much knowledge in ROS, Gazebo or in programming languages such as Python or C ++. The tool easies the understanding of quadrotor dynamics and related equations, as well as the development of control strategies. The present Chapter describes this user-friendly framework, the employed techniques are described in [14–17].

Section 2 presents a description of the employed model. Section 3 explains the optimal trajectory planning method. Section 4 describes the controllers available in the tool. Section 5 presents the graphical user interface developed by the tool. Section 6 presents the graphical and numerical results. In the end of the Chapter, the Section 7 emphasizes the main aspects and critical issues related.

2. Quadrotor model

The equations are described in the rigid body model of the quadrotor, and its displacement is related to an inertial *frame*, fixed on the Earth surface [18]. Three main *frames* are considered for a quadrotor model: Inertial (**I**), Vehicle (**V**) and Rigid Body (**B**), as illustrated in **Figure 1**.

The quadrotor owns 4 propellers in cross configuration. Each pair of propellers (1,3) and (2,4) rotates in opposite directions (**Figure 1**). Setting different rotor speeds to each pair ($\omega_2 \neq \omega_4$) or ($\omega_1 \neq \omega_3$) produces *roll* or *pitch* rotations with corresponding lateral motion. *Yaw* rotation results from rolling moments difference ($M_1 + M_3 - M_2 - M_4$) between propellers [9]. Maximum and minimum motor speeds are some of the parameters to be set in the model.

Newton-Euler equations describe quadrotor dynamics and kinematics [9, 18]. The Rotation Matrix represents the rotation around the three axes in the sequence $Z - X - Y$ (1), and is described in (2).

$$\mathbf{R}^{\mathbf{I}}_{\mathbf{B}_{\mathbf{Z-X-Y}}} = \mathbf{R}^{\mathbf{I}}_{\mathbf{A2}}(\psi)\mathbf{R}^{\mathbf{A2}}_{\mathbf{A1}}(\phi)\mathbf{R}^{\mathbf{A1}}_{\mathbf{B}}(\theta) \tag{1}$$

$$\mathbf{R}^{\mathbf{I}}_{\mathbf{B}_{\mathbf{Z-X-Y}}} \triangleq \begin{bmatrix} c\psi c\theta - s\phi s\psi s\theta & -c\phi s\psi & c\psi s\theta + c\theta s\phi s\psi \\ c\theta s\psi + c\psi s\phi s\theta & c\phi c\psi & s\psi s\theta - c\theta s\phi c\psi \\ -c\phi s\theta & s\phi & c\phi c\theta \end{bmatrix} \tag{2}$$

The angle ϕ around the x axis is the *roll angle*, the angle θ around the y axis is *pitch angle*, and the ψ angle around the z axis is *yaw angle*.

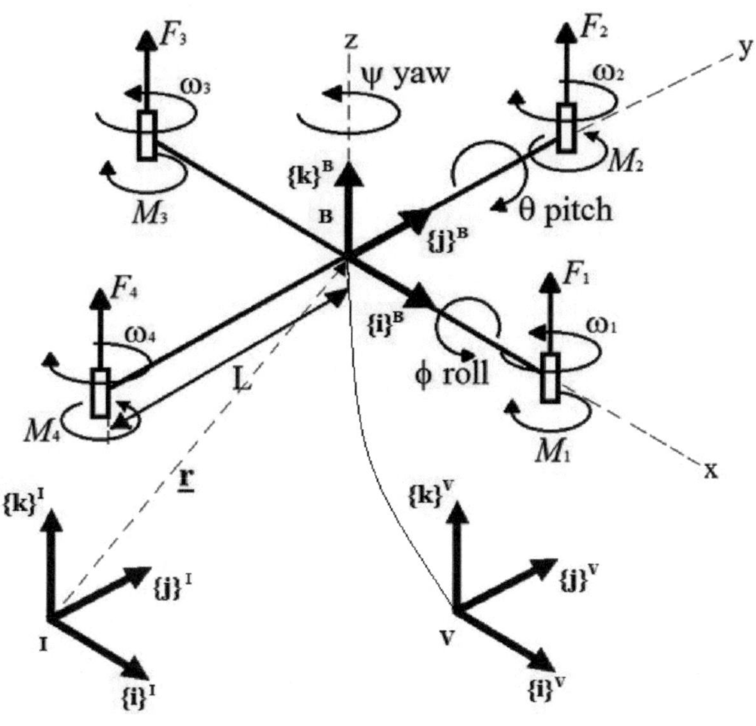

Figure 1.
Main frames in a quadrotor.

The orientation vector is defined by η and the position of quadrotor's center of mass is defined in the reference frame **I** by **r**. Angular velocities in the body frame **B** (p, q and r) are defined by ν (3).

$$\eta = \begin{bmatrix} \phi \\ \theta \\ \psi \end{bmatrix} \qquad \mathbf{r} = \begin{bmatrix} x \\ y \\ z \end{bmatrix} \qquad \nu = \begin{bmatrix} p \\ q \\ r \end{bmatrix} \tag{3}$$

The derivative of the angles ϕ, θ and ψ and the angular velocities measured by a sensor fixed to the *Frame* of the Rigid Body are not the same. p, q and r are the angular velocities around the x, y and z axes of the Rigid Body *Frame*. The relationship with the angular rates $\dot{\phi}$, $\dot{\theta}$ and $\dot{\psi}$ in the same *frame* **B** is in (4).

$$\begin{bmatrix} p \\ q \\ r \end{bmatrix} = \mathbf{T} \begin{bmatrix} \dot{\phi} \\ \dot{\theta} \\ \dot{\psi} \end{bmatrix} = \begin{bmatrix} c\theta & 0 & -c\phi s\theta \\ 0 & 1 & s\phi \\ s\theta & 0 & c\phi c\theta \end{bmatrix} \begin{bmatrix} \dot{\phi} \\ \dot{\theta} \\ \dot{\psi} \end{bmatrix} \tag{4}$$

Considering that the body is symmetrical with respect to the x-z and y-z planes of the *frame* **B**,and that the only forces acting on it are the weight and the four thrusts, its resulting linear acceleration with respect to the inertial *Frame* can be described using Newton's Second Law. Air drag forces are represented by a matrix in (5), and the values of A_x, A_y and A_z, are inputs on the parameters set.

$$m\{\ddot{\mathbf{r}}\}^{\mathrm{I}} = \left\{ \begin{bmatrix} 0 \\ 0 \\ -mg \end{bmatrix} \right\}^{\mathrm{I}} + \mathbf{R}_{\mathbf{B}}^{\mathrm{I}} \left\{ \begin{bmatrix} 0 \\ 0 \\ \sum F_i \end{bmatrix} - \begin{bmatrix} A_x & 0 & 0 \\ 0 & A_y & 0 \\ 0 & 0 & A_z \end{bmatrix} \right\}^{\mathrm{B}} \tag{5}$$

Besides the force each rotor produces a moment in the rigid body perpendicular to the plane of rotation of the propeller (M_i), contrary to the direction of rotation of the propellers. L is the size of the quadrotor arm (**Figure 1**) and **J** the inertia matrix presented in (6).

$$\mathbf{J} = \begin{bmatrix} J_{xx} & 0 & 0 \\ 0 & J_{yy} & 0 \\ 0 & 0 & J_{zz} \end{bmatrix} \tag{6}$$

Due to the forces produced by the rotors, moments are produced in the rigid body ($L \cdot F_i$), causing the system to rotate around the x and y axes. The rotation around the z axis is due to the torque created by the rotation of the motors, which are fixed to the plant. Based on the Coriolis Equation, the equation that describes the angular acceleration in the *Frame* **B** is in (7).

$$\mathbf{J}\left\{\begin{bmatrix} \dot{p} \\ \dot{q} \\ \dot{r} \end{bmatrix}\right\}^{B} = \left\{\begin{bmatrix} L(F_2 - F_4) \\ L(F_1 - F_3) \\ M_1 + M_2 + M_3 + M_4 \end{bmatrix}\right\}^{B} - \left\{\begin{bmatrix} p \\ q \\ r \end{bmatrix}\right\}^{B} \times \mathbf{J}\left\{\begin{bmatrix} p \\ q \\ r \end{bmatrix}\right\}^{B} \tag{7}$$

For simulating the measurement noise, the tool allows to add a random signal to position and orientation variables and their derivatives. A first order delay between the comands u_1 and $\mathbf{u_2}$ and rotor speed variations, with a time constant τ seconds, was included in the tool to simulate the rotor dynamics.

3. Trajectory planning for a set of waypoints

Due to the low inertia of quadrotor it is necessary to calculate a smooth trajectory to minimize the risk of collapse. Euler–Lagrange equations are used to find the minimum *snap* trajectory [15, 19]. For a two waypoints trajectory, the boundary conditions of position, velocity, acceleration, and *jerk* are defined in **Table 1**.

With this boundary conditions the equations' coefficients for the two-points optimal desired trajectory are calculated for each coordinate of position (x_{des}, y_{des} and z_{des}) and orientation (ψ_{des}) (8).

$$\begin{bmatrix} x_{des} \\ y_{des} \\ z_{des} \\ \psi_{des} \end{bmatrix} = \begin{bmatrix} c_{1,7} & c_{1,6} & c_{1,5} & c_{1,4} & c_{1,3} & c_{1,2} & c_{1,1} & c_{1,0} \\ c_{2,7} & c_{2,6} & c_{2,5} & c_{2,4} & c_{2,3} & c_{2,2} & c_{2,1} & c_{2,0} \\ c_{3,7} & c_{3,6} & c_{3,5} & c_{3,4} & c_{3,3} & c_{3,2} & c_{3,1} & c_{3,0} \\ c_{4,7} & c_{4,6} & c_{4,5} & c_{4,4} & c_{4,3} & c_{4,2} & c_{4,1} & c_{4,0} \end{bmatrix} \begin{bmatrix} t^7 \\ t^6 \\ t^5 \\ t^4 \\ t^3 \\ t^2 \\ t \\ 1 \end{bmatrix} \tag{8}$$

Time t	**Position $x_{des}(t)$**	**Velocity $\dot{x}_{des}(t)$**	**Acceleration $\ddot{x}_{des}(t)$**	**Jerk $\dddot{x}_{des}(t)$**
0	$x_{des}(0)$	0	0	0
T	$x_{des}(T)$	0	0	0

Table 1.
Boundary conditions.

Desired angles for *roll* (ϕ_{des}) and *pitch* (θ_{des}) are calculated from ψ_{des}, the equations depend on which control strategy is employed.

This procedure yields a minimum *snap* trajectory for two points. For additional waypoints it is necessary to consider more equations and intermediary restrictions [15].

The tool calculates the complete optimized trajectory for any set of waypoints. Desired trajectory $\mathbf{r}_{des}$, orientation ψ_{des}, and their derivatives are calculated for being used in the control algorithms.

4. Attitude and trajectory following control strategies

Some of the challenges to be overcome in quadrotor operation are: attitude stability for large angles, trajectory following, collision avoidance through aggressive maneuvers, monitoring, and others [2, 4].

The control architecture employed in the following control strategies uses two cascaded loops. The inner loop (attitude control) runs in a fast time-scale and is assumed exponentially stable. The outer loop (position control) runs in a slow time-scale, with a higher bandwidth [4]. All of them employ a feed-forward with proportional plus derivative structure (FF-PD). The tool is open to easily add more control strategies.

4.1 Small angle control

Small Angle control assumes an operation not too far from the hovering condition. A simple heuristic method with FF-PD control calculates the required accelerations for the desired trajectory (9).

$$\begin{bmatrix} \ddot{x}_c \\ \ddot{y}_c \\ \ddot{z}_c \end{bmatrix} = \mathbf{r}_{des} + K_p\,\mathbf{e}_r + K_d\,\mathbf{e}_r \tag{9}$$

It is assumed small deviations from zero in *roll* and *pitch* angles, small deviations in the *yaw* angle from the desired value, and angular velocities close to zero. The algorithm assumes all upward-pointing thrust vectors (control signal u_1 in (10)).

$$u_1 = m(g + \ddot{z}_c) \tag{10}$$

After linear simplifications the equations for attitude control are defined in (11) and (12) [16].

$$\eta_c = \begin{bmatrix} \phi_c \\ \theta_c \\ \psi_c \end{bmatrix} = \begin{bmatrix} \frac{1}{g}\left(\ddot{x}_c \sin(\psi_{des}) - \ddot{y}_c \cos(\psi_{des})\right) \\ \frac{1}{g}\left(\ddot{x}_c \cos(\psi_{des}) + \ddot{y}_c \sin(\psi_{des})\right) \\ \psi_{des} \end{bmatrix} \tag{11}$$

$$\nu_c = \begin{bmatrix} p_c \\ q_c \\ r_c \end{bmatrix} = \mathbf{T}\eta_c \tag{12}$$

A PD control law is used for attitude control and to calculate $\mathbf{u}_2$ (13) [16].

$$\mathbf{u}_2 = K_R(\eta_c - \eta) + K_\nu(\nu_c - \nu) \tag{13}$$

4.2 Desired thrust vector control

At high speeds and *roll* and *pitch* angles far away from zero, a more robust strategy is necessary. A FF-PD control law calculates the control signal u_1 for the trajectory following and compensates the gravity. Vector $\mathbf{t}$ is calculated with the attitude for the desired effect (14). As u_1 acts along the z direction in frame $\mathbf{B}$ (axis $\mathbf{b}_z$), it must be referred to frame $\mathbf{I}$ (15).

$$\mathbf{t} = m\left(\mathbf{r}_{des} + K_p\,\mathbf{e}_r + K_d\,\mathbf{e}_r\right) + mg\mathbf{a}_z \tag{14}$$

$$u_1 = \mathbf{t}^T\mathbf{R}\,\mathbf{b}_z \tag{15}$$

The axis $\mathbf{b}_z$ is desired to be aligned with $\mathbf{t}$. The desired rotation matrix $\mathbf{R}_{des}$ is calculated from the equation of rotation of $\mathbf{b}_z$ in the direction of $\mathbf{t}$ (16).

$$\mathbf{R}_{des}\mathbf{b}_z = \frac{\mathbf{t}}{\|\mathbf{t}\|} \tag{16}$$

As we know ψ_{des} we use (16) to calculate ϕ_{des} and θ_{des}. $\mathbf{R}_{des}$ is constructed with these three angles with the same structure of (2). The error in rotation $\Delta\mathbf{R}$ is calculated from (17).

$$\Delta\mathbf{R} = \left(\mathbf{R}_{\mathbf{B}}^{\mathbf{I}}\right)^T\mathbf{R}_{des} \tag{17}$$

With the Rodrigues formula [20], the axis of rotation $\mathbf{v}$ and the rotation angle β are calculated (18). $\mathbf{I}_{3\times3}$ is the identity matrix and $\hat{\mathbf{v}}$ is the skew-symmetric matrix of $\mathbf{v}$ (19). The related error is calculated with (20).

$$\Delta\mathbf{R} = \mathbf{I}_{3\times3}\cos\beta + \mathbf{v}\mathbf{v}^T(1 - \cos\beta) + \hat{\mathbf{v}}\sin\beta \tag{18}$$

$$\hat{\mathbf{v}} = \begin{bmatrix} 0 & -v_3 & v_2 \\ v_3 & 0 & -v_1 \\ -v_2 & v_1 & 0 \end{bmatrix} \tag{19}$$

$$\mathbf{e}_R = \beta\mathbf{v} \tag{20}$$

A PD control law is used to calculated $\mathbf{u}_2$ (22).

$$\mathbf{e}_\nu = \nu_c - \nu \tag{21}$$

$$\mathbf{u}_2 = \nu \times \mathbf{J}\nu + \mathbf{J}(-K_R\,\mathbf{e}_R - K_\nu\,\mathbf{e}_\nu) \tag{22}$$

Basin of attraction Ψ limits the set of rotations from which the quadrotor is able to converge to the hovering state. It is a dimension of the set of angular and linear velocities for a stable performance. For this control strategy it must be lower than 2 (23).

$$\Psi = \mathrm{tr}\left(\mathbf{I}_{3\times3} - \mathbf{R}_{des}^T\,\mathbf{R}_{\mathbf{B}}^{\mathbf{I}}\right) < 2 \tag{23}$$

4.3 Geometric tracking control

Geometric Tracking Control exhibits almost global exponential attractiveness to the zero equilibrium of tracking errors [17]. $\mathbf{t}$ and u_1 are calculated as in (14) and (15). $\mathbf{b}_x$ is the desired direction vector in the first body-fixed axis (24).

$$\mathbf{b}_x = \begin{bmatrix} \cos\psi_{des} \\ \sin\psi_{des} \\ 0 \end{bmatrix} \tag{24}$$

With **t** and $\mathbf{b}_x$ is possible to calculate $\mathbf{R}_{des}$ with (25) to (27) [17].

$$\mathbf{b}_z = \frac{\mathbf{t}}{\|\mathbf{t}\|} \tag{25}$$

$$\mathbf{b}_y = \frac{\mathbf{b}_z \times \mathbf{b}_x}{\|\mathbf{b}_z \times \mathbf{b}_x\|} \tag{26}$$

$$\mathbf{R}_{des} = \begin{bmatrix} \mathbf{b}_y \times \mathbf{b}_z & \mathbf{b}_y & \mathbf{b}_z \end{bmatrix} \tag{27}$$

(27) calculates the desired attitude for the quadrotor given **t** and ψ_{des}. The basin of attraction Ψ (28) is bigger than in the previous strategy (23) as this is a more robust approach [17].

$$\Psi = \frac{1}{2}\operatorname{tr}\left(\mathbf{I}_{3\times3} - \mathbf{R}_{des}^T \mathbf{R}_{\mathbf{B}}^{\mathrm{I}}\right) < 2 \tag{28}$$

Attitude tracking error and angular velocity error are calculated from (29, 30). The control vector $\mathbf{u}_2$ is calculated in (31) [17].

$$\mathbf{e}_R = \frac{1}{2}\left(\mathbf{R}_{des}^T \mathbf{R}_{\mathbf{B}}^{\mathrm{I}} - {\mathbf{R}_{\mathbf{B}}^{\mathrm{I}}}^T \mathbf{R}_{des}\right)^{\vee} \tag{29}$$

$$\mathbf{e}_\nu = \nu - {\mathbf{R}_{\mathbf{B}}^{\mathrm{I}}}^T \mathbf{R}_{des}\,\nu_{des} \tag{30}$$

$$\mathbf{u}_2 = \nu \times \mathbf{J}\nu + \mathbf{J}(-K_R\mathbf{e}_R - K_\nu\mathbf{e}_\nu) - \mathbf{J}\left(\hat{\nu}\,{\mathbf{R}_{\mathbf{B}}^{\mathrm{I}}}^T \mathbf{R}_{des}\,\nu_{des} - {\mathbf{R}_{\mathbf{B}}^{\mathrm{I}}}^T \mathbf{R}_{des}\,\dot{\nu}_{des}\right) \tag{31}$$

4.4 Particle swarm optimization for control gains tuning

A PSO algorithm is employed to tune the control gains. Some adjustments were performed to reduce the processing [14, 15], s. Each particle α_i is a vector with the proportional and derivative gains (32).

$$\alpha_i = \begin{bmatrix} K_p & K_d & K_R & K_\nu \end{bmatrix} \tag{32}$$

Using a predefined set of waypoints chosen by the user, the code calculates the desired trajectory and tests each particle, calculating the RMS error. The PSO algorithm evolves in the direction of the best validated solution in each iteration, until achieving a minimum error tolerance. For faster convergence, every time a particle is evaluated, the evaluation is interrupted in the middle of the trajectory if the error increases above a predefined limit.

5. The graphical user interface

A user-friendly 3D animated GUI was developed in MatLab. It is able to evaluate and compare the performance of various quadrotor control strategies for any user-chosen trajectory. A few of quadrotor parameters are easily set in this interface [15].

A red vertical line indicates the upper side of the device, and a red circle the front rotor. Waypoints are represented by red markers, the desired trajectory is on

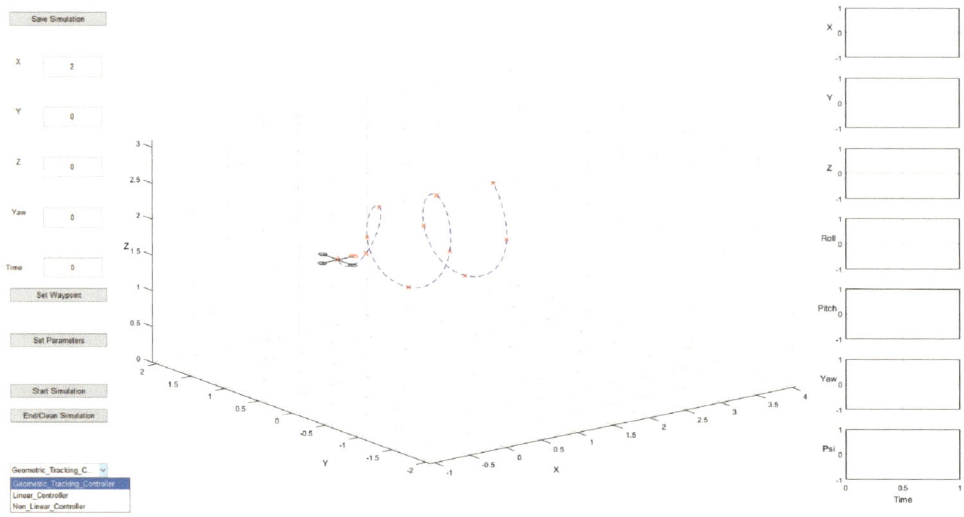

Figure 2.
Graphical user interface.

dashed blue line, and the 3D interface axis limits are automatically adjusted from the waypoints (**Figure 2**).

Waypoints, quadrotor parameters and simulation comands are set in th buttons on the left side of the GUI.

The user may test its own control strategy just by creating the corresponding code like the following example:

```
function New_Controller
% declare global variables
global quad;
% holds the quadrotor in the last waypoint
if(quad.iteracao > length(quad.rdes(1,:)))
Controlador_Position_Hold()
end
% start the controller code and calculate the global control commands
quad.u1=...
quad.u2=...
```

The code `New_Controller.m` must be stored in the folder `Controllers`, and will be recognized in the dropdown menu in the lower left side of **Figure 2**.

6. Results

The tool developed in MatLab and is available in GitHub [21]. The values employed for these results are in **Table 2** [3].

For simulating the measurement noise, a random signal was added to position ($\pm 0.01\, m$) and orientation ($\pm 0.5°$) variables and their derivatives. Maximum and minimum motor speeds make the simulation more reliable. A time constant τ of 0.1 s for dynamic rotor speed variation was considered.

Two sets of waypoints of challenging trajectories were tested, each one was simulated in three different initial conditions: Case 1 starting with an orientation

Parameter	Value	Units
J_{xx}	4.856×10^{-3}	$kg.m^2$
J_{yy}	4.856×10^{-3}	$kg.m^2$
J_{zz}	8.801×10^{-3}	$kg.m^2$
k	2.980×10^{-6}	$N.s^2/rad^2$
g	9.81	m/s^2
m	0.468	kg
A_x	0.25	kg/s
A_y	0.25	kg/s
A_z	0.25	kg/s
b	1.100×10^{-7}	$N.m.s^2/rad^2$
L	0.225	m
$N_{i\,max}$	8500	rpm
$N_{i\,min}$	1300	rpm
τ	0.1	s

Table 2.
Quadrotor parameters [3].

angle of $\phi(0) = 0°$, Case 2 with $\phi(0) = 88°$, and Case 3 with $\phi(0) = 178°$. The three control strategies available in the tool were graphically and numerically validated.

6.1 Elliptical helix trajectory

The first trajectory waypoints (**Table 3**) describe an elliptical helix in a forward trajectory along with the x axis of inertial frame $\mathbf{i}_x$.

$t(s)$	$x_{\text{des}}(m)$	$y_{\text{des}}(m)$	$z_{\text{des}}(m)$	$\psi_{\text{des}}(\text{rad})$
0.00	−0.40	0.00	2.00	0.00
1.20	0.00	0.00	2.00	0.00
1.80	0.20	0.00	2.60	0.00
2.40	0.40	0.40	2.00	1.57
3.00	0.60	0.00	1.40	3.14
3.60	0.80	−0.40	2.00	4.71
4.20	1.00	0.00	2.60	6.28
4.80	1.20	0.40	2.00	7.85
5.40	1.40	0.00	1.40	9.42
6.00	1.60	−0.40	2.00	11.00
6.60	1.80	0.00	2.60	12.57
7.20	1.80	0.00	2.60	12.57

Table 3.
Waypoints test 1.

With this set it was calculated the desired optimal trajectory. Using this trajectory the PSO algorithm calculates the optimal gains for each of the three evaluated control algorithms.

With this gains the simulation was performed and validation errors in each of the four coordinates were registered. **Tables 4–6** present the gains used in each case, and corresponding validation errors.

When calculating the gains using PSO it was observed that the processing time increases substantially with the initial orientation angle ($\phi(0)$). It was also observed in cases 2 and 3 that once a set of optimal gains is calculated, later results do not reduce substantially the validation error.

Small angle control is more sensible to variations in rotational positions as observed in the results. Optimal small angle control gains are bigger than rotational (**Tables 4–6**). The opposite occurs for DTV and geometric tracking.

Case 3 is the most challenging since the quadrotor starts almost upside down. Geometric tracking controller is more sensitive to noise error, PSO did not succeed in finding a set of gains in Case 3.

A visual validation is possible using the tool. Graphics in **Figure 3** show the performance of small angle control in Case 3. The quadrotor drops and recoverers the vertical position, and continues along with the desired trajectory. Graphics on the right supports a visual validation of position and orientation variables.

In the graphics of propellers' speeds of **Figure 4**, dashed red lines indicate the speed extremes. In extreme situations, the controller attains the propellers' limits.

Figure 5 presents the performance of DTV Control on Case 3. It is observed a faster reaction and a more accurate trajectory following.

Near the third point the trajectory leads the quadrotor to its limits. Propellers' speeds in **Figure 6** show a more stable performance than the previous strategy.

Control Strategy	K_p	K_d	K_R	K_ν	x_v %	y_v %	z_v %	ψ_v %
Small Angle	17.96	7.11	0.231	0.16	1.373	7.763	6.301	1.0300
DTV	18.91	10.77	44.920	23.43	2.112	4.907	5.300	0.6455
Geometric Tracking	7.34	7.35	90.590	20.75	3.172	8.807	26.680	0.3973

Table 4.
Elliptical helix trajectory case 1, $\phi(0) = 0°$.

Control Strategy	K_p	K_d	K_R	K_ν	x_v %	y_v %	z_v %	ψ_v %
Small Angle	22.76	2.00	0.765	0.197	2.01	32.69	26.22	0.6344
DTV	19.70	4.85	155.400	26.710	1.58	22.03	27.65	0.4949
Geometric Tracking	8.24	5.62	220.200	27.670	3.17	35.94	51.57	0.4027

Table 5.
Elliptical helix trajectory case 2, $\phi(0) = 88°$.

Control Strategy	K_p	K_d	K_R	K_ν	x_v %	y_v %	z_v %	ψ_v %
Small Angle	8.180	2.349	2.45	0.4371	5.988	64.98	146.1	1.223
DTV	7.268	3.434	236.40	31.4300	4.709	33.71	105.4	1.351
Geometric Tracking	—	—	—	—	—	—	—	—

Table 6.
Elliptical helix trajectory case 3, $\phi(0) = 178°$.

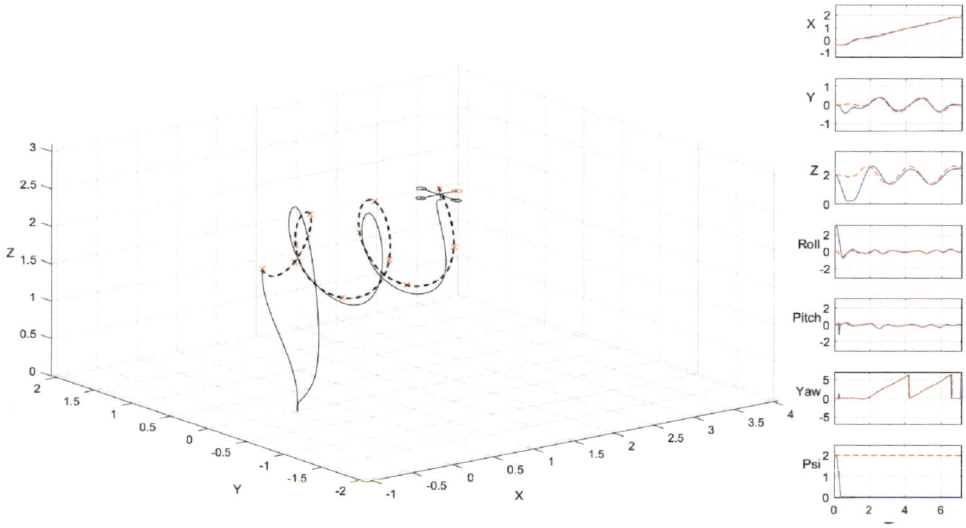

Figure 3.
Elliptical helix trajectory case 3 small angle control, $\phi(0) = 178°$.

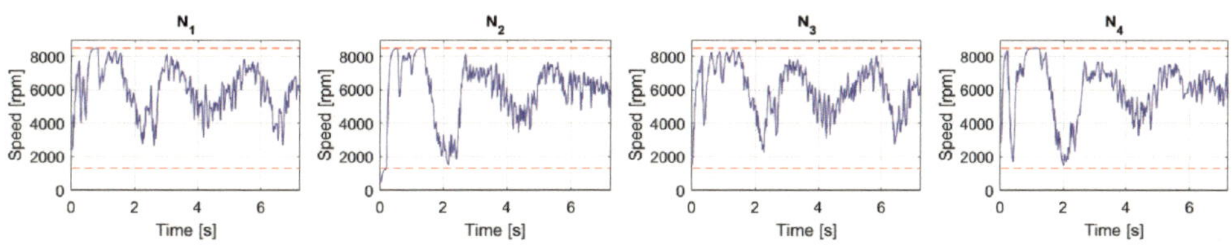

Figure 4.
Propellers' speeds in rpm with small angle control.

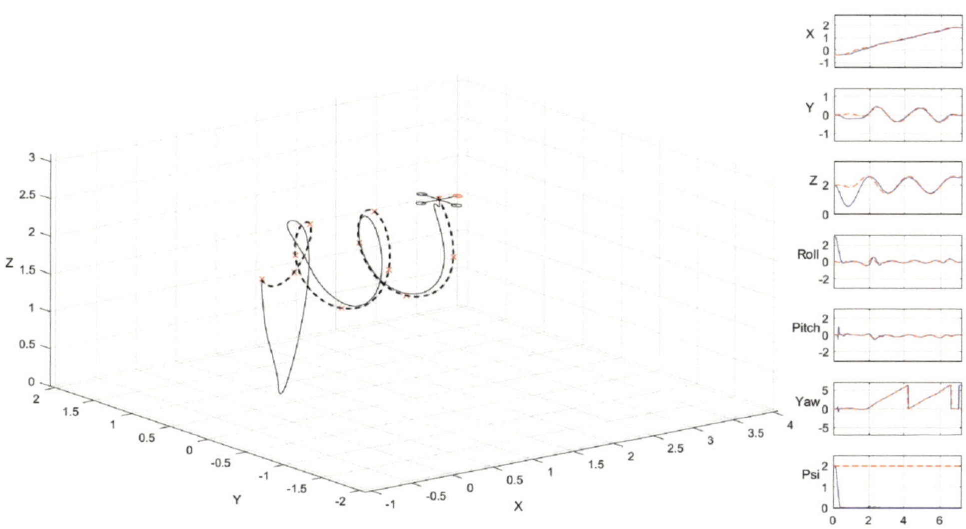

Figure 5.
Elliptical Helix Trajectory Case 3 DTV Control, $\phi(0) = 178°$.

Compared with small angle, this strategy seems to be more robust and accurate.

Figure 7 presents the performance of geometric tracking control in Case 2. This strategy reacts faster to challenging situations, since it recovers in a shorter

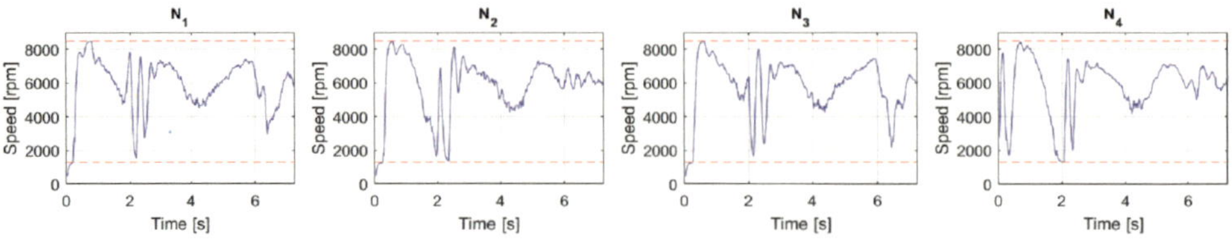

Figure 6.
Propellers' speeds in rpm with DTV control.

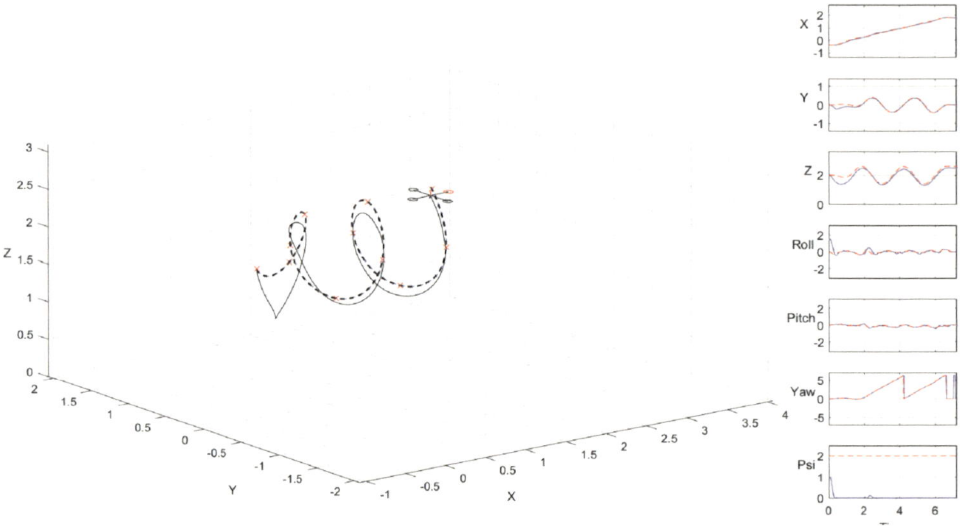

Figure 7.
Elliptical helix trajectory case 2 geometric tracking, $\phi(0) = 88°$.

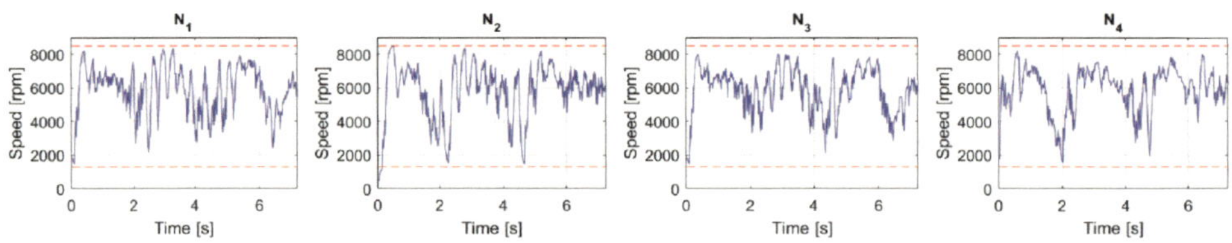

Figure 8.
Propellers' speeds in rpm with geometric tracking.

time its hovering orientation. This fast reaction makes it more sensitive to measuring noise.

Propellers' speeds in **Figure 8** confirm the faster command reaction of the controller compared with **Figures 4** and **6**. This strategy is also sensitive to nonlinear behaviour. When it reaches the propeller limits, it is highly prone to instability.

The gains calculated for the more challenging Cases were later tested on Cases 1 and 2. Despite being less optimal they displayed a stable behaviour. The performance is presented in **Table** 7.

Graphical results of small angle control performance on Case 1, when using the optimal gains compared with the performance with the gains calculated for Case 3. The system holds the attitude performance. The same comparison was performed for DTV control. The decrease in performance is lower than with the small angle controller [15].

Control Strategy	$\phi(0)^{\circ}$	x_v %	y_v %	z_v %	ψ_v %
Small Angle	0°	1.373	7.763	6.301	1.0300
Small Angle[1]	0°	3.111	21.960	28.390	0.2430
Small Angle	88°	2.010	32.690	26.220	0.6344
Small Angle[1]	88°	3.242	51.760	49.580	0.3378
DTV	0°	2.112	4.907	5.300	0.6455
DTV[1]	0°	4.625	15.650	13.310	0.2350
DTV	88°	1.584	22.030	27.650	0.4949
DTV[1]	88°	4.886	41.580	44.530	0.3509
Geometric Tracking	0°	3.172	8.807	26.680	0.3973
Geometric Tracking[2]	0°	3.338	5.917	28.660	0.2615

[1] *Calculated with Gains of Case 3.*
[2] *Calculated with Gains of Case 2.*

Table 7.
Validation comparison for cases 1 and 2.

6.2 Lemniscate Shape Trajectory

A second set of waypoints presented in **Table 8**, depicts a lemniscate shape trajectory varying the position in Z, orienting the front of quadrotor (*yaw* angle ϕ) in the direction of displacement.

Figure 9 depicts the second trajectory.

$t(s)$	$x_{\text{des}}(m)$	$y_{\text{des}}(m)$	$z_{\text{des}}(m)$	$\psi_{\text{des}}(\text{rad})$
0.00	0.00	0.00	2.00	−1.57
0.40	0.12	−0.28	1.58	−0.79
0.80	0.40	−0.40	1.40	0.00
1.20	0.68	−0.28	1.58	0.79
1.60	0.97	0.00	2.00	0.79
2.00	1.25	0.28	2.42	0.79
2.40	1.53	0.40	2.60	0.00
2.80	1.81	0.28	2.42	−0.79
3.20	1.93	0.00	2.00	−1.57
3.60	1.81	−0.28	1.58	−2.36
4.00	1.53	−0.40	1.40	−3.14
4.40	1.25	−0.28	1.58	−3.93
4.80	0.97	0.00	2.00	−3.93
5.20	0.68	0.28	2.42	−3.93
6.60	0.40	0.40	2.60	−3.14
6.00	0.12	0.28	2.42	−2.36
6.40	0.00	0.00	2.00	−1.57

Table 8.
Waypoints test 2.

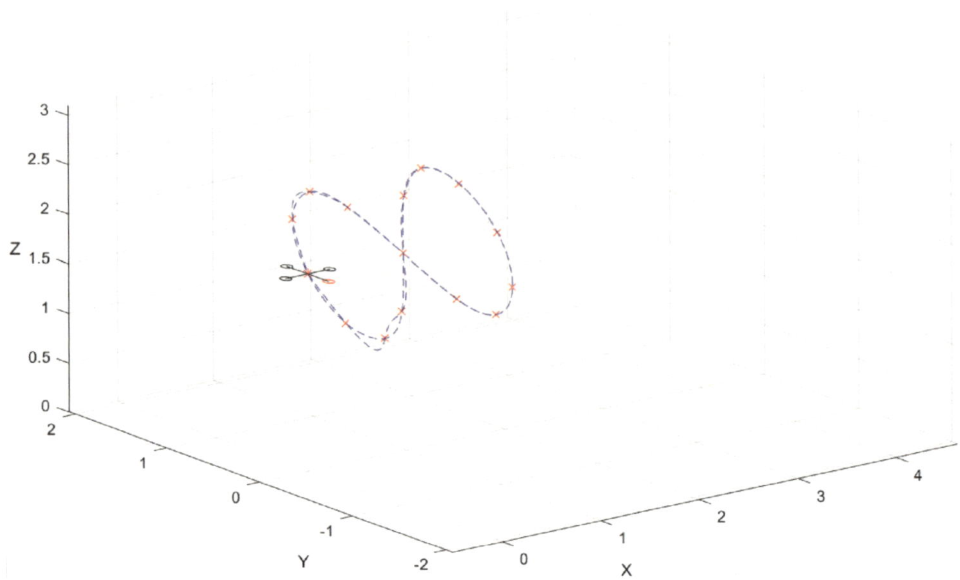

Figure 9.
Lemniscate shape trajectory.

As with the to previous trajectory, the small angle control compensates its limitations with lower gains to guarantee stability, gains are bigger for position control than for attitude.

DTV Control is again the best validated strategy. Geometric tracking presents a lower performance due to its sensitivity to noise and nonlinear operation conditions. The DTV control had presented the best validation even in the most challenging conditions.

Detailed results and analysis of this trajectory are available in [15].

Using the tool the graphical and numerical analysis was easier. Not much programming knowledge was necessary for using and configuration, just basic MatLab programming. Access to quadrotor parameters of the analyzed device, analyzing the influence of measurements noise, and comparing and simulating different control strategies is easier than with other visual platforms. Graphical and numerical simulation results are easily available with the interface buttons.

7. Summary

Quadrotor control is a fascinating research area, but the equations involved and programming skills requirements can be arduous for initiating students. It is a worth to develop motivational appliances for beginners. This was the motivation to present a beginner-friendly visual interface tool for the development of quadrotor control strategies. It is easy to understand, device characteristics are simple to configure, and control algorithms can be implemented and analyzed with not much effort. It is not necessary to have a deep knowledge in programming languages, and may be an introduction to this field of research.

This tool uses RMS and basin of attraction for numerical validation, and the GUI may help to evaluate stability, robustness, and accuracy. It integrates these criteria in a unique interface and helps to measure and visualize details and requirements that may not be so clear using other visual tools.

Baseline controllers are offered so students may compare performance, and is open to easily introduce other strategies for comparison. A trajectory planning based on minimum *snap* give support to trajectory following control, and GUI allows the evaluation in dimensions of position and time.

The tool makes easier and faster to realize critical quadrotor requirements and limitations for challenging applications. These requirements may be related with the complexity of a defined trajectory, the weakness of a control strategy, or the improvements that may be carried out in the quadrotor (size, weight, propeller power, etc.) to accomplish the desired results.

Other controllers can be studied and compared using this tool, such as *Backstepping* and intelligent strategies. In the trajectory planning stage, applications with obstacles may be simulated. Support for multiple quadrotors, communication with the controller via Robotics Operating System (ROS) and implementation of obstacles are some of the future improvements planned for this tool.

Abbreviations

ROS	*Robot Operating System*
PSO	*Particle Swarm Optimization*
DTV	*Desired Thrust Vector*
RMS	*Root Mean Square*
GUI	*Graphical User Interface*
UAV	*Unmanned Aerial Vehicles*
PD	*Proportional plus Derivative Control*
PID	*Proportional Integral Derivative Control*
LQR	*Linear Quadratic Regulator*
LQG	*Linear Quadratic Gaussian*
FF-PD	*Feed-forward Proportional Derivative Control*

Nomenclature

I	Inertial frame
V	Vehicle frame
B	Rigid Body frame
ω_i	Speed of the rotor i
M_i	Moment produced by the rotor i
F_i	Propulsion force produced by the rotor i
ϕ	*Roll*
θ	*Pitch*
ψ	*Yaw*
$\dot{\phi}$	Rate of change of *Roll*
$\dot{\theta}$	Rate of change of *Pitch*
$\dot{\psi}$	Rate of change of *Yaw*
$s\phi$,$s\theta$,$s\psi$	Sine of angles ϕ, θ and ψ
$c\phi$,$c\theta$,$c\psi$	Cosine of angles ϕ, θ and ψ
$\mathbf{R}_{\mathrm{X}}^{\mathrm{Y}}$	Rotation matrix of a vector represented in an arbitrary *frame* X for an arbitrary Y *frame*
p	Angular speed related to the x axis in the *Frame* **B**
q	Angular speed related to the y axis in the *Frame* **B**
r	Angular speed related to the z axis in the *Frame* **B**

$\dot{p}$	Angular acceleration related to the x axis in the *Frame* **B**
$\dot{q}$	Angular acceleration related to the y axis in the *Frame* **B**
$\dot{r}$	Angular acceleration related to the z axis in the *Frame* **B**
η	Angular position vector in the *Frame* **I**
$\mathbf{r}$	Linear position vector in the *Frame* **I**
$\ddot{\mathbf{r}}$	Linear acceleration vector in the *Frame* **I**
ν	Angular speed vector in the *Frame* **B**
$\mathbf{T}$	Rotation matrix for angular velocity
L	Size of the quadrotor arm
$\mathbf{J}$	Inertia matrix
x	Linear position of quadrotor in the x axis of *frame* I
y	Linear position of quadrotor in the y axis of *frame* I
z	Linear position of quadrotor in the z axis of *frame* I
$\mathbf{e_r}$	Linear position error vector
$\mathbf{e_r}$	Linear velocity error vector
K_p	Linear control gain
K_d	Derivative control gain
m	Quadrotor mass
g	Gravity constant
K_R	Angular position control gain
K_ν	Angular velocity control gain
$\mathbf{t}$	Desired orientation vector
$\mathbf{b_i}$	Desired direction vector on axis i of *frame* B
$\mathbf{I_{3x3}}$	Identity matrix
$\mathbf{v}$	Axis of rotation of Rodriguez formula
β	Angle of rotation of Rodriguez formula
$\mathbf{e_R}$	Angular position error vector
$\mathbf{e}_\nu$	Angular velocity error vector
Ψ	Basin of attraction
α	Particle of the PSO algorithm
k	Constant of the rotor force
b	Constant of the rotor moment
τ	Time constant for rotor delay dynamics

Author details

Manuel A. Rendón[1,2]

1 Federal University of Juiz de Fora - UFJF, Juiz de Fora, MG, Brazil

2 Electromechanical Energy Conversion Group - GCEME, Juiz de Fora, Brazil

*Address all correspondence to: manuel.rendon@ufjf.edu.br

IntechOpen

© 2021 The Author(s). Licensee IntechOpen. This chapter is distributed under the terms of the Creative Commons Attribution License (http://creativecommons.org/licenses/by/3.0), which permits unrestricted use, distribution, and reproduction in any medium, provided the original work is properly cited. CC BY

References

[1] Annaz F. Uav testbed training platform development using panda3d. Industrial Robot: An International Journal. 2015;**42**(5):450-456

[2] A Zul Azfar and D Hazry. Simple gui design for monitoring of a remotely operated quadrotor unmanned aerial vehicle (uav). In *2011 IEEE 7th International Colloquium on Signal Processing and its Applications*, pages 23–27. IEEE, 2011.

[3] A Tayebi and S McGilvray. Attitude stabilization of a four-rotor aerial robot. In Decision and Control, 2004. CDC. *43rd IEEE Conference on*, volume 2, pages 1216–1221. IEEE, 2004.

[4] Tiago P Nascimento and Martin Saska. Position and attitude control of multi-rotor aerial vehicles: A survey. *Annual Reviews in Control*, 2019.

[5] Can Dikmen I, Arisoy A, Hakan Temeltas. Attitude control of a quadrotor. In Recent Advances in Space Technologies. *RAST'09*. In: *4th International Conference on*, pages 722–727. IEEE. *2009*. p. 2009

[6] Gabriel M Hoffmann, Haomiao Huang, Steven L Waslander, and Claire J Tomlin. Quadrotor helicopter flight dynamics and control: Theory and experiment. In *Proc. of the AIAA Guidance, Navigation, and Control Conference*, volume 2, page 4, 2007.

[7] Domingues JMB. Quadrotor prototype. *Dissertacio*: *Uneversidade Tecnica deLisboa*; 2009

[8] Ly Dat Minh and Cheolkeun Ha. Modeling and control of quadrotor mav using vision-based measurement. In *Strategic Technology (IFOST), 2010 International Forum on*, pages 70–75. IEEE, 2010.

[9] Samir Bouabdallah and Roland Siegwart. Backstepping and sliding-mode techniques applied to an indoor micro quadrotor. In *Proceedings of the 2005 IEEE international conference on robotics and automation*, pages 2247–2252. IEEE, 2005. ISBN 078038914X.

[10] C Nicol, CJB Macnab, and A Ramirez-Serrano. Robust neural network control of a quadrotor helicopter. In *Electrical and Computer Engineering, 2008. CCECE 2008. Canadian Conference on*, pages 001233–001238. IEEE, 2008.

[11] Steven Lake Waslander, Gabriel M Hoffmann, Jung Soon Jang, and Claire J Tomlin. Multi-agent quadrotor testbed control design: Integral sliding mode vs. reinforcement learning. In *Intelligent Robots and Systems, 2005.(IROS 2005). 2005 IEEE/RSJ International Conference on*, pages 3712–3717. IEEE, 2005.

[12] Fadri Furrer, Michael Burri, Markus Achtelik, and Roland Siegwart. Rotors—a modular gazebo mav simulator framework. In *Robot Operating System (ROS)*, pages 595–625. Springer, 2016.

[13] Subhan Khan, Mujtaba Hussain Jaffery, Athar Hanif, and Muhammad Rizwan Asif. Teaching tool for a control systems laboratory using a quadrotor as a plant in matlab. *IEEE Transactions on Education*, 60(4): 249–256, 2017.

[14] Manuel A Rendón and Felipe F Martins. Unmanned quadrotor path following nonlinear control tuning using particle swarm optimization. In *2018 Latin American Robotic Symposium, LARC 2018 and 2018 Workshop on Robotics in Education WRE 2018*, pages 509–514. IEEE, 2018.

[15] Manuel A Rendón, Felipe F Martins, and Luis Gustavo Ganimi. A visual interface tool for development of quadrotor control strategies. *Journal of Intelligent & Robotic Systems*, pages 1–18, 2020.

[16] Mellinger D, Michael N, Kumar V. Trajectory generation and control for precise aggressive maneuvers with quadrotors. The International Journal of Robotics Research. 2012;**31**(5):664-674

[17] Lee T, Leoky M, N Harris McClamroch. Geometric tracking control of a quadrotor uav on se (3). In Decision and Control (CDC). *49th IEEE Conference on*, pages 5420–5425. IEEE. *2010*;**2010**

[18] Randal Beard. Quadrotor dynamics and control rev 0.1. Technical document, Brigham Young University, Provo, UT, USA, February 2008.

[19] Daniel Mellinger and Vijay Kumar. Minimum snap trajectory generation and control for quadrotors. In *2011 IEEE International Conference on Robotics and Automation*, pages 2520–2525. IEEE, 2011.

[20] Jian S Dai. Euler–Rodrigues formula variations, quaternion conjugation and intrinsic connections. Mechanism and Machine Theory, 92: 144–152, 2015.

[21] Felipe F. Martins and Manuel A. Rendón. Easyquadsim: Easy quadrotor simulator. GitHub, 2019. URL EasyQuadSim.

Chapter 8

AI-Based Approach for Lawn Length Estimation in Robotic Lawn Mowers

Yoichi Shiraishi, Haohao Zhang and Kazuhiro Motegi

Abstract

This chapter describes a part of autonomous driving of work vehicles. This type of autonomous driving consists of work sensing and mobility control. Particularly, this chapter focuses on autonomous work sensing and mobility control of a commercial electric robotic lawn mower, and proposes an AI-based approach for work vehicles such as a robotic lawn mower. These two functions, work sensing and mobililty control, have a close correlation. In terms of efficiency, the traveling speed of a lawn mower, for example, should be reduced when the workload is high, and vice versa. At the same time, it is important to conserve the battery that is used for both work execution and mobility. Based on these requirements, this chapter is focused on developing an estimation system for estimating lawn grass lengths or ground conditions in a robotic lawn mower. To this end, two AI algorithms, namely, random forest (RF) and shallow neural network (SNN), are developed and evaluated on observation data obtained by a fusion of ten types of sensor data. The RF algorithm evaluated on data from the fusion of sensors achieved 92.3% correct estimation ratio in several experiments on real-world lawn grass areas, while the SNN achieved 95.0%. Furthermore, the accuracy of the SNN is 94.0% in experiments where sensor data are continuously obtained while the robotic lawn mower is operating. Presently, the proposed estimation system is being developed by integrating two motor control systems into a robotic lawn mower, one for lawn grass cutting and the other for the robot's mobility.

Keywords: AI, robotic lawn mower, work vehicle, Random Forests, Neural Network, embedded system, Hybrid Twin

1. Introduction

Recently, automated driving algorithms and systems for work vehicles such as robotic lawn grass or grass mowers (robo-mowers) [1–3], autonomous snow blowers [4], automatic guided vehicles (AGVs) [5], autonomous delivery vehicles [6], and autonomous mobile robots (AMRs) [7], have attracted much attention. These vehicles are made possible by significant advances in sensor fusion technology, high-performance embedded systems, AI algorithms and advanced model-based design or development methods. Particularly, Industry 4.0 or Society 5.0 needs the digital transformation or smart factory in industry and, now, AGVs and AMRs perform some tasks that are essential for constructing the automatic

 IntechOpen

production lines. As these vehicles share a battery for their work and mobility, the interaction between their functions should be effectively controlled to reduce battery charging frequency and time, as well as working time. The authors in [4] attempted to optimize the skidding control of a snow blower, which has a motor for mobility and an engine for blowing snow. In this case, the engine is also used to generate electric power, which is then used to charge the battery that powers the motor. In this sense, its work and mobility share the energy, thus the two functions require effective energy management. Precise control handling of work load in these work vehicles is critical for optimizing its energy management.

In the following sections, because commercial robo-mowers [8] are popular and readily available for experiments, a robo-mower is used as an example for optimizing energy management in work vehicles. Mobility control is the next research theme for optimizing the energy management of robo-mowers. Current robo-mowers do not recognize the length of lawn grasses or ground conditions such as dirt, gravel, or concrete. As a result, the motor for cutting lawn grasses operates at a constant rotation speed from start to finish. Therefore, if the rotation speed of the motor for a lawn grass cutter is precisely controlled, battery wastage can be avoided. Moreover, because the control of grass cutting and mobility is correlated, the mobility speed should be controlled according to the lawn grass lengths and ground conditions. Then, the working time can also be reduced. Therefore, the precise estimation of lawn grass lengths using effective sensor data is required in the first stage, i.e., preventing battery wastage. Then, in the second step, the mobility of robo-mowers is controlled according to the estimation results from the first stage. Finally, a cooperative control of a group of robo-mowers is researched [3] and implemented. In particular, the group control of robo-mowers becomes meaningful when the performance of each robo-mower is optimized.

In this study, an AI-based approach is adopted for the estimation of lawn grass lengths from the fusion of sensor data. A random forest (RF) algorithm and shallow neural network (SNN) are suggested. Ten measurement data types are obtained from sensors attached to a robo-mower. The combination of sensor data types is essential for lawn grass estimation, that is, a sensor fusion problem is discussed. In general, the sensor fusion and use of big data have attracted many researchers' interest. Recently, there have been detailed surveys on the combination of sensor fusion and big data analysis [9, 10]. Some applications to actual problems have also been reported [11–13]. The popular approach for big data analysis is the use of machine learning. Takami G., et al. [11] studied the observation of plant status. They used three kinds of sensors and a deep learning (DL) algorithm for big data analysis. The details of the DL are not described, and the processing time of the observation system is not known; however, it may be useful to learn that they predicted the deterioration of sensors performance through their combination. Alonso S., et al. [12] also adopted the same approach for observing a screw compressor in a chiller. They used five kinds of sensor data and a 1D convolutional neural network (CNN) for their analysis. The adoption of 1D CNN makes monitoring faster and real-time processing is realized. Their approach is probably suitable for data without any estimated features; however, in this study some features may be efficient for the estimation task in advance. Li C., et al. [13] performed the diagnosis of rotating machinery. They used vibration sensor signals, and the Gaussian-Bernoulli deep Boltzmann machine was used for their analysis. The accuracy of fault estimation was evaluated; however, its real-time processing requirement was not mentioned. Therefore, this approach cannot be applied to the problem dealt with in the following.

In the experiments of the proposed AI-based approach, the application of RF algorithm to the fusion of seven sensors attained a 92.3% correct estimation ratio in several experiments on actual lawn grass areas. In addition, the application of SNN attained 95.0%. Moreover, the accuracy of SNN is 94.0% in such trials in which sensing data are continuously obtained while the robo-mower is in operation. The proposed estimation system is being developed by integrating it with two types of motor control systems for grass cutting and robot mobility, respectively. At present, the authors are promoting the research and development of robo-mowers for commercial use by collaborating with an automobile company.

The outline of this chapter is as follows. Section 2 describes the Hybrid Twin ™, which is the basic idea for controlling the robo-mower in real time. Section 3 describes robo-mower used in this chapter; however, the discussions are not limited to this robo-mower. Moreover, the estimation problem of lawn grass lengths is also defined in this section. Section 4 describes the proposed RF and SNN algorithms. Section 5 describes the experimental results based on the big data obtained from sensor fusion and a set of features for classifying the sensor data are. Furthermore, the set of necessary sensors and performance evaluations of the proposed algorithms are stated. Section 6 describes the evaluation of the proposed SNN algorithm when applied to the consecutive sensor data obtained in real-world use. Finally, the chapter is summarized in Section 7.

2. Hybrid twin within work vehicle

The Hybrid Twin ™ approach [14] is efficient for real-time object control. The proposed estimation method is useful for controlling the operation of lawn grass cutter motor and a mobility motor. When the robot is operating in an area with long lawn grasses, the motor should be set to the maximum rotation speed. On the other hand, the rotation speed should be reduced or stopped when the robo-mower is operating in an area with short lawn grasses or in an area without any lawn grasses, respectively. As a result, battery consumption will decrease. Furthermore, if it is possible to control the robo-mower's speed so that it decreases or increases according to the length of lawn grasses, the working time will be greatly reduced. When a ground without lawn grasses is identified, laying the electric cable that defines the boundary of the area is no more necessary and, as a result, the required maintenance is reduced.

Digital Twin has become popular for implementing smart factory, and it has been used [15] for controlling mission-critical systems, such as nuclear plant, airplane control, or rocket control in the aerospace industry. The Digital Twin constructed in the virtual space means a twin of a real space object. The twin is a precise model, and its behaviors are reproduced in the virtual space. The Hybrid Twin ™ is an extension of Digital Twin. As the target system has become large and complicated, the Virtual Twin has been separated from the Digital Twin, as shown in **Figure 1**.

The Digital Twin only obtains data from a fusion of sensors, and measurement data with some abstractions are transferred to the Virtual Twin. The Virtual Twin is a precise co-simulator consisting of subsystems obtained using a model-based design method. This simulator must be sufficiently fast, and it is usually a 1-D simulator, which is a high-speed version of the original 3-D simulator is used. The Hybrid is a combination of the Virtual Twin and Digital Twin, and the optimized state of the real system on time $(t + \Delta t)$ must be fedback to the real system from the state on time (t). This loop is iterated over with the time interval (Δt). As a result, the state of the real system is optimized in real time. Measurement data are extracted from the fusion of sensors for robo-mower operations, and noise reduction is applied to the obtained parameters in the Digital Twin. This means that the Digital Twin is an accurate numerical model of real objects. The Virtual Twin receives the obtained data s(t) at

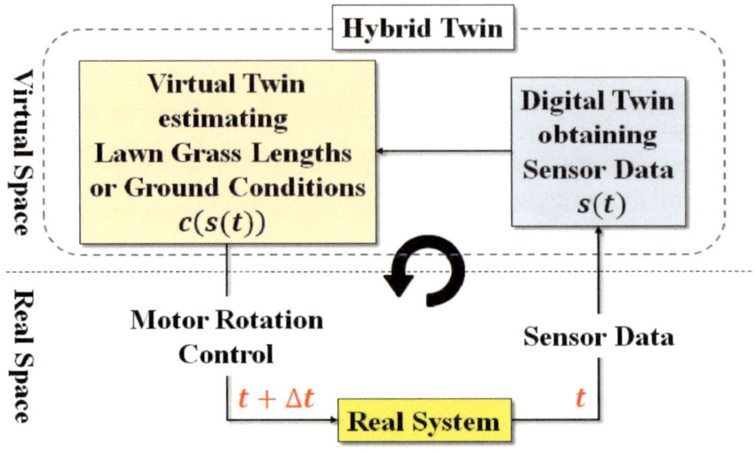

Figure 1.
Hybrid twin within work vehicles.

time (t), and then, estimation results $c(s(t))$ are obtained using AI algorithms. The control parameter set at time $(t + \Delta t)$ for motors is given to the real robo-mower. This loop is repeated during the operation of the robo-mower.

3. Work vehicles

This section describes the work vehicle used in the following discussions and experiments. A commercial robo-mower [8] is used as the experimental hardware for evaluating the proposed algorithms with the available fusion of sensors.

3.1 Robotic Lawn mower

The exterior and the backside of the robo-mower are shown in **Figure 2(a)**. The system configuration is shown in **Figure 2(b)**. It has two kinds of motor controllers,

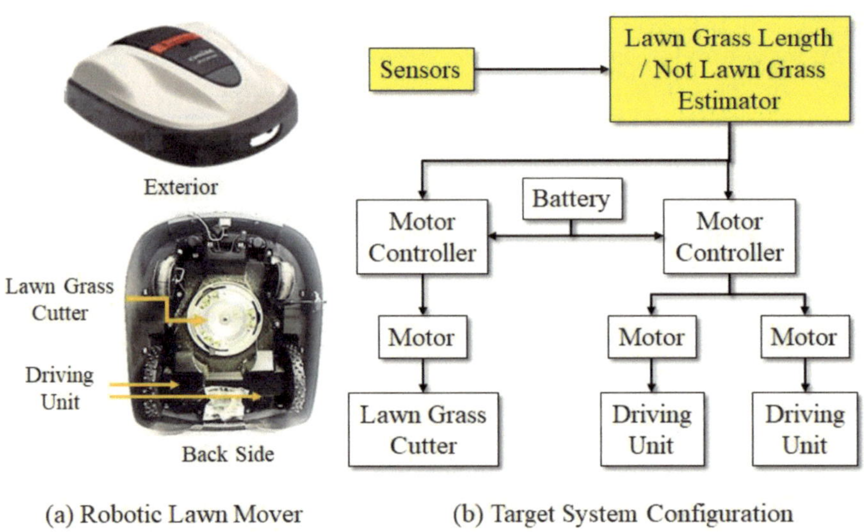

Figure 2.
Robotic Lawn mower and target system configuration.

one of which is used for controlling the lawn grass cutter and another is for its mobility. These controllers are powered by the same battery, and therefore, battery wastage depends on their usages. The subsystem, "Lawn Grass Length/Not Lawn Grass Estimator" is newly developed.

The robo-mower is used for experiments by attaching sensors, a single-board computer, a personal computer and peripheral devices on the robo-mower, as shown in **Figure 3**. All these devices are managed by an ROS (robot operating system) running on the personal computer. The robo-mower can be autonomously driven; however, it is controlled using a Bluetooth controlling device in the experiments to increase the accuracy of the experiments. A camera can be used as a sensor, but it is inadequate for the experiments due to its high cost of image processing software and hardware. In the experiments, it will be shown that no camera is needed for the required estimation.

3.2 Fusion of sensors

The sensors attached to the robo-mower shown in **Figure 3** are listed in **Table 1**. The robo-mower has originally been equipped with built-in sensors. The 9-axis

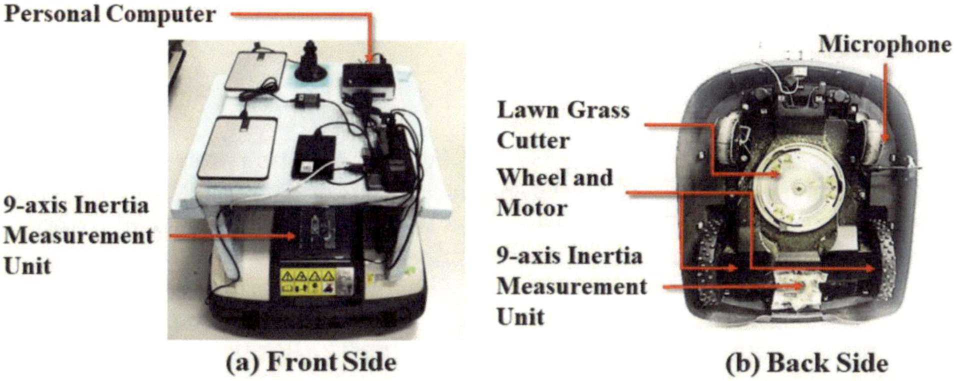

Figure 3.
Robotic Lawn mower with sensors and devices for development.

Sensors	Mounting positions	Measurements
9-axis Inertial Measurement Unit	Inside of the Robo-mower	Acceleration
		Angular Acceleration
	Surface of the Robo-mower	Acceleration
		Angular Acceleration
Built-In	Battery	Voltage
		Current
		Power
	Rotation of Grass Cutting Motor	
	Rotation of Traveling Motor	
	Horizontal/Vertical Acceleration	

Table 1.
Fusion of sensors.

inertial measurement units (IMUs), MPU-9250 [16], are attached inside and to the surface of the robo-mower to measure acceleration and angular acceleration. Six built-in sensors are available for measuring the corresponding parameters as shown in **Table 1**. Here the noise of sensors is negligibly small; however, outliers are excluded in the Digital Twin. The adequate fusion of these sensors is determined in each of the proposed algorithms, and this is verified in the experiments.

3.3 Estimation problem of Lawn grass lengths

The estimation problem of lawn grass lengths and ground conditions is defined below. The problem is to estimate lawn grass lengths in real time using sensor fusion data. The objective function is to increase the accuracy of estimation.

The Estimation Problem.

Input: set of available sensors, robo-mower's specifications, set of areas labeled.

long lawn grass, short lawn grass, and without lawn. grasses, some of which are specified as test areas.

Output: fusion of sensors necessary for estimation and estimation results for test areas.

Objective Function: maximization of estimation accuracy.

4. AI-based approach

An AI-based approach is adopted for solving the estimation problem. The reason for this is that a combination of different types of sensor data should be handled, and the definition of long or short lawn grass is determined by the height of the lawn grass cutter from the ground. Moreover, a human operator estimates the length of a lawn grass based on sounds made by the lawn grass cutter while cutting grasses. Estimation using an AI-based approach is expected to be more efficient and accurate than estimation based on human judgment. The RF algorithm and SNN are adopted considering the execution speed in real-world applications.

4.1 Random Forest algorithm

The RF algorithm, a machine learning algorithm, originates from Breiman [17], and recently, its deep version has also been proposed [18]. This algorithm is used for classification, regression or clustering, etc. and is a type of ensemble algorithm using a set of decision trees as weak learners to avoid over-fitting and to maintain its high generalization performance. It is fast and achieves a comparatively high performance. According to the study [18], the deep RF algorithm achieves better results in specific applications while performing nearly as well in other wide applications. In the following, a specific RF algorithm is developed.

An RF algorithm consists of a given number of binary decision trees. The training and inference phases of the algorithm are shown in **Figure 4(a)** and **(b)**, respectively. In the configuration of binary decision trees, a set of training data sampled from input data is given to each of the binary decision trees. Then, the binary decision trees are constructed, as shown in **Figure 5**.

The data consist of the followings:

$\{n_i\}$ $(i = 1, 2, \cdots, p)$: input data for classification, regression or clustering, etc.

$\{x_i\}$ $(i = 1, 2, \cdots, q)$: features for classifying input data $\{n_i\}$.

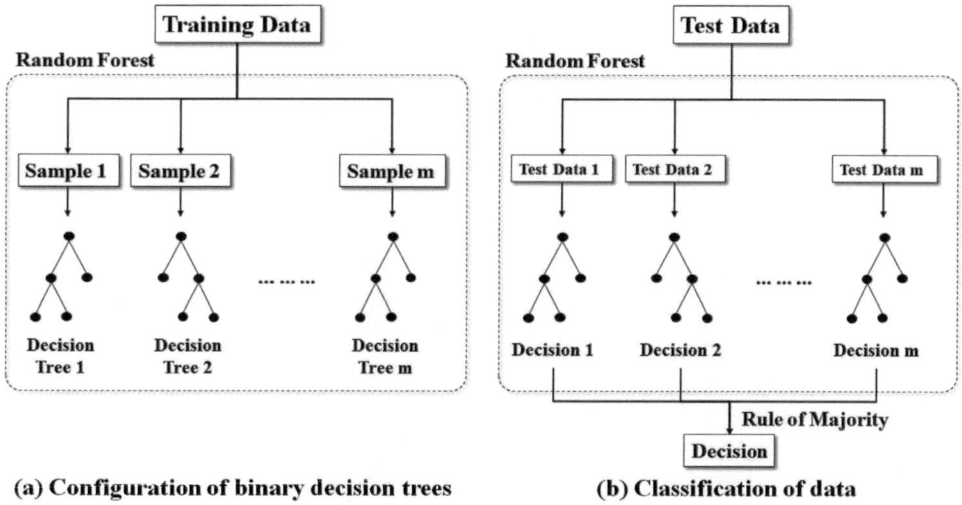

Figure 4.
Random forest algorithm.

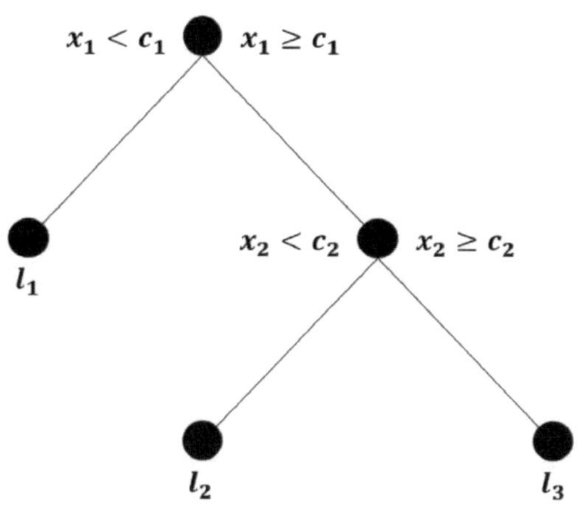

Figure 5.
Construction of binary decision tree.

4.1.1 Configuration of binary decision tree

An example of a binary decision tree is shown in **Figure 5**. In the root node, the input data are divided into two subsets using the conditions, $x_1 < c_1$ and $x_1 \geq c_1$. If the data satisfy the condition $x_1 < c_1$, the data are classified into the class l_1 as shown in this figure. When all data are classified into the corresponding classes (that is, leaves), the binary decision tree is completed. Here, for example, a classification and regression tree algorithm is used for classification, and the objective function is Gini's diversity index [18]. All parameters in binary decision trees are used in the classification phase.

4.1.2 Classification of data

For example, Bagging, an ensemble algorithm, is used for data classification. In this case, data that should be classified are distributed to all binary decision trees, and the decision of each binary decision tree is obtained. The final decision, that is, the class to which the data belong, is determined on the basis of the majority rule. This process is faster if the binary decision trees are executed in parallel, and the decision quality is higher than if only one binary decision tree is used. As a 1D simulator, or a Virtual Twin, processing time for estimating the target area is essential.

4.2 Shallow neural network

As another AI-based approach, a SNN is used. "Shallow" means it has only one hidden layer, and this is expected to fasten processing. The estimation performance of SNN is compared with that of the RF algorithm.

4.2.1 Design of shallow neural network

The deep neural network performs well in image recognition, and it has been used in autonomous driving of automobiles, appearance inspection, image recognition of robots, and other applications. However, it requires a large amount of training data, and accordingly, a huge amount of processing time is needed in network training. In addition, the inference should also be executed on a GPU machine. On the contrary, the size of signal data is not so big because they are time-series, and no deep neural network may be needed for its recognition. For example, the on-line hammering sound inspection based on the simplest neural network with no hidden layers, a support vector machine, achieves more than 99% accuracy within a short time [19].

In the following, an SNN, as shown in **Figure 6**, is constructed. Here, the number of hidden layers is only one, and this layer has ten neurons. The number of neurons in the input layer equals the size of input signal length or statistical features such as the maximum value, minimum value, average value, median value, standard deviation value, peakedness value, and skewness value, all of which are obtained from the input sensor signal. The number of neurons in the output layer is two, that is, areas with or without lawn grasses, or with short or long lawn grasses. A hyperbolic tangent activation function and a softmax function are incorporated into the output layer.

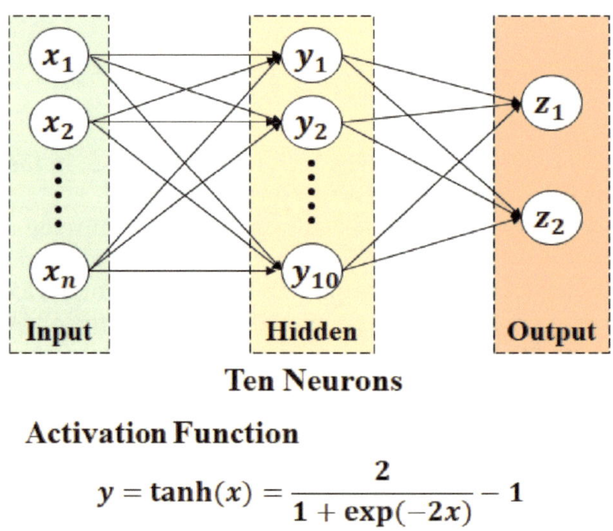

Figure 6.
Shallow neural network configuration.

4.2.2 Application of shallow neural network

In the following, two SNN models are developed with a cascade connection. The first model estimates the target areas that are with or without lawn grasses because this can be distinguished according to the acceleration of the robo-mower operating in a corresponding area. The second model estimates the height of the lawn grass as long or short. It is expected that this might be determined by checking the current of the motor depending on the load on the cutting blade. These two models are connected in series, that is, a cascade configuration. Furthermore, these SNNs are trained independently.

5. Experimental results

This section focuses on the experiments and evaluations of the RF algorithm and SNN on real-world sensing data.

5.1 Experiments on RF algorithm

5.1.1 Measurement data

The data measured by the sensors are obtained by driving the robo-mower on a field with long lawn grasses and short lawn grasses as well as without lawn grasses. The actual remote-controlled robo-mower is shown in **Figure 7**. The remote-control system through Bluetooth communication is incorporated in the robo-mower by mounting a mini-PC and running an ROS on it. The mini-PC can also handle the collected sensor data.

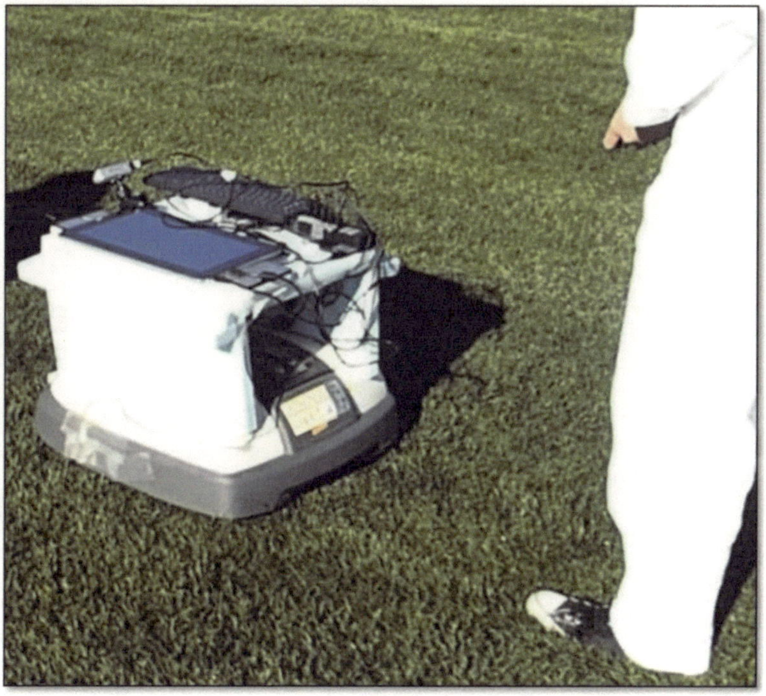

Figure 7.
Remote-controlled driving of Robo-mower.

The collected data are manually categorized into long and short lawn grass datasets according to the height of the grass cutter from the ground. If the length of lawn grasses exceeds the height of the grass cutter, the lawn grass length is defined as long and otherwise, short. When the heights of lawn grasses and grass cutter are equal, a human operator determines whether the lawn grass is long or short according to the operating sound of the grass cutter. The measurement data are collected within a total time of 2.3 h. All data are collected on flatlands on sunny days.

5.1.2 Features for classifying data

Statistical features of input data $\{x_i\}$ $(i = 1, 2, \cdots, q)$ for classification are calculated, including (i) maximum value, (ii) minimum value, (iii) average value, (iv) median value, (v) standard deviation value, (vi) kurtosis value, and (vii) skewness value. The values of seven feature types are normalized into the interval [-1,1]. These features are used in configuring binary decision trees, and they are calculated for each time frame obtained approximately every 3.2 s over 2.3-h measurement data. The details of the collected data are shown in **Table 2**. Even if the total time for data collection is less than 2.3 h because of, for example, some issues with measurement devices, the obtained data are used.

In the experiments, a subset of time frames obtained from each field data is used for configuring the binary decision trees, and the completed forest is applied to the remaining test data. Then, classification performance of the RF algorithm is evaluated.

The number of time frames (training data) used to configure the binary decision trees in each group is chosen at random from the measurement data. The remaining time frames are used as test data for evaluating the RF's performance. These are shown in **Table 3**.

5.1.3 Evaluation criteria

Each of the time frame data has its label, that is, long lawn grasses, short lawn grasses, and not lawn grasses, and the estimation can be verified. This process

Groups	Number of time frames
Long Lawn Grasses	2,356
Short Lawn Grasses	1,575
Not Lawn Grasses	3,374
Total	7,305

Table 2.
Specifications of measurement data for evaluating RF algorithm.

Groups	Number of time frames	
	For training	For testing
Long Lawn Grasses	1,686	670
Short Lawn Grasses	904	671
Not Lawn Grasses	2,705	669
Total	5,295	2,010

Table 3.
Number of time frames for training and testing RF algorithm.

consists of two stages. The first stage is used to estimate whether an area is with or without lawn grasses. In the second stage, an area is further estimated whether it has long or short lawn grasses when it is estimated to have lawn grasses. In the testing, four kinds of evaluation criteria are used. These are defined below [20].

		Actuals	
		Positive	**Negative**
Predictions	**Positive**	*TP*	*FP*
	Negative	*FN*	*TN*

1. Accuracy

$$Accuracy = \frac{TP + TN}{TP + FP + FN + TN} \tag{1}$$

2. Precision

$$Precision = \frac{TP}{TP + FP} \tag{2}$$

3. Recall

$$Recall = \frac{TP}{TP + FN} \tag{3}$$

4. F-Measure

$$F - Measure = \frac{2 * Presision * Recall}{Precision + Recall} \tag{4}$$

where $TP, TN, FP,$ and FN denote “True Positive,” “True Negative,” “False Positive,” and “False Negative,” respectively.

5.1.4 Evaluation results

Seven combinations of sensor data used are shown in **Table 4**. These combinations cover all possible cases. Using the measurement data from **C1** to **C7**, the best combination of sensor data is determined based on the above-mentioned evaluation criteria.

The procedure of experiments is as follows.

1. Select the sensor data corresponding to the cases shown in **Table 4** collected in three ground conditions, that is, “Long Lawn Grasses,” “Short Lawn Grasses,” and “Not Lawn Grasses.”

Determine the subset of sensor data (1) and partition it to configure the binary decision trees and to test the RF algorithm according to the number of the time frames shown in **Table 3**.

2. Configure the binary decision trees.

3. Evaluate the performances of the RF algorithm based on the evaluation criteria.

The number of binary decision trees, that is, the size of the forest is set to 1,000. Each binary decision tree is configured using the seven features mentioned in 5.1.2 until each leaf coincides with one of three ground conditions. An example of an actually constructed binary decision tree is shown in **Figure 8**. Here, the feature, median value, obtained from built-in vertical angle sensor data with its threshold 370.75, is used for classifying the input data on the root node. The class

Sensors	Mounting positions	Measurements	C1	C2	C3	C4	C5	C6	C7
9-axis Inertial Measurement Unit	Inside of Body	Acceleration	√	√		√	√	√	√
		Angular Acceleration	√	√		√	√	√	√
	Surface of Body	Acceleration		√			√		√
		Angular Acceleration		√			√		√
Built-In	Battery	Voltage			√	√	√	√	√
		Current			√	√	√	√	√
		Power			√	√	√	√	√
	Rotation of Grass Cutting Motor				√	√	√	√	√
	Rotation of Traveling Motor				√	√	√	√	√
	Horizontal/Vertical Angles				√	√	√		

Table 4.
Combinations of sensor data.

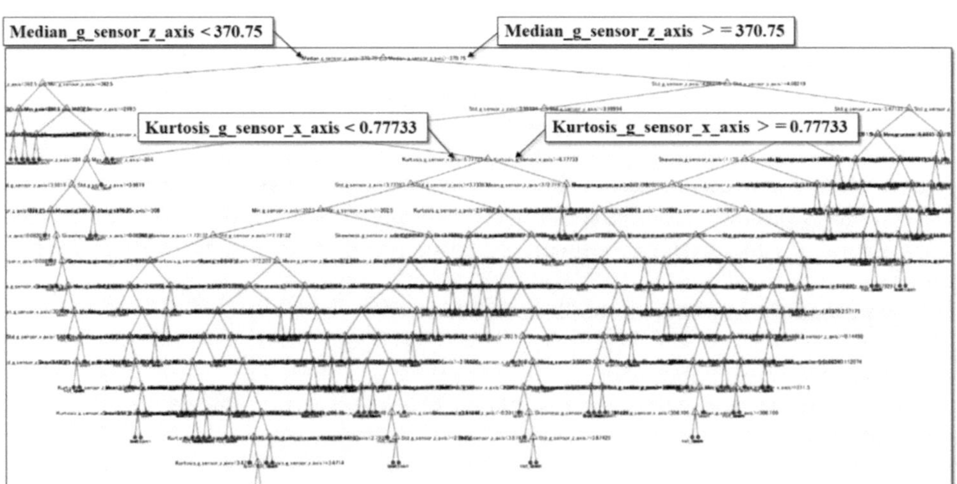

Figure 8.
Example of binary decision tree in RF algorithm.

"TreeBagger" included in "Statics and Machine Learning Toolbox" in MATLAB [21] is used for implementing the RF algorithm. The processing time for configuring 1,000 binary decision trees is less than ten minutes on a PC with the standard performances. The completed forest is applied to the testing data whose size is approximately 700 in each of the three ground conditions shown in **Table 3**. The estimating time is negligibly small, and this is no issue in the actual Hybrid Twin approach. **Table 5** shows the performance of the algorithm according to different sensor data combinations. Seven cases are evaluated with respect to the measurement criteria in each ground condition. The most important performance is the accuracy, and it increases when the built-in sensor data are used. Particularly, **C6** and **C7**, excluding the built-in horizontal or vertical angle sensor, have higher accuracy. It seems reasonable that the battery status and motor rotation conditions contribute to higher performance because the rotation of the motor becomes high when it encounters long lawn grasses. On the other hand, the load on both the grass cutting motor and traveling motor is reduced when the robo-mower travels on a ground without lawn grasses. From the evaluation results, **C6** is desirable among seven cases. The reason is that

1. the accuracy is high, with a difference of only 0.1 points from maximum 92.28%,

2. the recall ratio of Short Lawn Grasses, 87.08%, is the highest.

Especially, the low recall ratio of Short Lawn Grasses means that the probability of incorrectly recognizing short lawn grasses as long lawn grasses or a ground other than lawn grasses is high. Then, the traveling speed of the robo-mower is reduced, and the rotation of the grass cutting motor is increased. This would increase the working time and waste electric energy. Moreover, when a short lawn grass area is incorrectly classified as a ground without lawn grasses, the robo-mower will not move on the areas and will not cut lawn grasses. Therefore, it would be concluded that **C6** is the best combination in this evaluation results. Sensor data, including the acceleration and angular acceleration values obtained using the 9-axis IMU attached inside the robo-mower; the voltage, current, and power of the battery; the rotation

Combinations of sensor data		C1	C2	C3	C4	C5	C6	C7
Accuracy		75.84	77.52	86.32	90.92	91.58	92.18	92.28
Long Lawn Grasses	**Precision**	77.06	70.70	92.68	92.60	91.46	92.66	91.04
	Recall	73.02	67.70	93.46	93.60	93.64	93.44	93.56
	F-measure	74.99	69.17	93.07	93.10	92.54	93.05	92.28
Short Lawn Grasses	**Precision**	72.66	73.84	87.30	90.48	92.12	89.40	92.04
	Recall	60.88	68.28	70.06	82.00	83.26	87.08	85.58
	F-measure	66.25	70.95	77.73	86.03	87.47	88.22	88.69
Not Lawn Grasses	**Precision**	77.06	86.34	80.24	89.70	91.34	94.40	93.72
	Recall	93.60	96.60	95.46	97.20	97.90	96.08	97.70
	F-measure	84.53	91.18	87.19	93.30	94.51	95.23	95.67
Average of Precision, Recalls, and F-measures		75.56	77.20	86.35	90.89	91.58	92.17	92.25

Table 5.
Evaluation results for sensor fusions.

speed of the grass cutting and traveling motors obtained using built-in sensors, are used in **C6**.

5.2 Experiments for SNN

5.2.1 Available sensors

The built-in sensors attached to the robo-mower are listed in **Table 1**. Six built-in sensors are available for measuring the corresponding parameters shown in **Table 1**. The objectives of the experiments are first to evaluate the accuracy of lawn grass height estimation and ground condition estimation and, second, to compare the SNN's results with those of the RF algorithm.

5.2.2 Measurement data

The data measured by sensors are collected while driving the robo-mower on a field with long and short lawn grasses as well as without lawn grasses. The three types of lawn grasses are shown in **Figure 9(a)–(c)** are used for the experiments. In each of these cases, the long lawn grass case and short lawn grass case are performed by adjusting the lawn grass cutting blade height from the ground. Similarly, several ground conditions without lawn grasses are adopted as shown in **Figure 9(d)–(f)**, which are asphalt, gravel, and stone pavement, respectively.

The collected data are categorized into three groups. The first group is for long lawn grasses, that is, the height of lawn grass is larger than the specified one. The second group is for the short lawn grasses, that is, the lawn grass is shorter than or equal to the specified one. The third group is the field without lawn grasses, that is, dirt, gravel, stone pavement, tiled, asphalt, or concrete fields. The measurement data are collected for a total driving time of 2.3 h for each group with various fields.

Figure 9.
Variations of target areas.

5.2.3 Fusion of sensors for SNN

The fusions of sensors for two SNNs SNN1 and SNN2 are experimentally determined. Ten types of sensors, as shown in **Table 1**. As **Figure 10** shows, SNN1 and SNN2 are cascade connected, where SNN1 estimates if a target area is with or without lawn grasses, and SNN2 provides the result that an area is with long/short lawn grasses. The basic idea behind this configuration is that sensor fusion may differ with these two types of estimations.

For SNN1, the Horizontal/Vertical Acceleration sensor seems adequate, and the combinations of x, y, and z-axis sensor data are experimented with and evaluated on some datasets. The obtained accuracy is shown in **Figure 10**. For the estimation of lawn lengths, sensor fusion indicating the load of the lawn grass cutter seems effective. Therefore, the measurement data obtained from the combinations of the battery sensor and duty ratio given to the cutting motor are evaluated with some datasets. As shown in **Figure 10**, the multiplication of battery voltage and the duty ratio for the cutting motor achieves maximum accuracy. The differences are minor; however, this multiplication is probably the reason for showing the load of grass cutting motor.

As a result, the x-axis and z-axis values obtained from the Horizontal/Vertical Acceleration sensor are used as inputs for SNN1. Similarly, the multiplication of battery voltage and the duty ratio for the grass cutting motor, respectively, obtained from Battery Voltage and Rotation of Grass Cutting Motor are used as inputs for SNN2.

5.2.4 Features for classifying data

The input data features are (i) maximum value, (ii) minimum value, (iii) average value, (iv) median value, (v) standard deviation value, (vi) peakedness value, and (vii) skewness value. In signal recognition based on machine learning, some

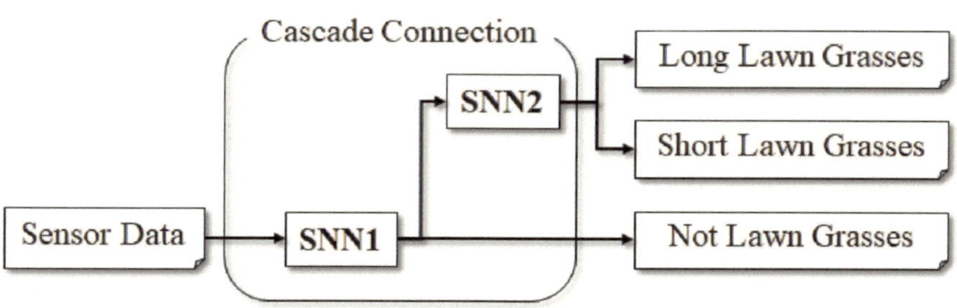

Sensor Data for SNN1

	Horizontal / Vertical Acceleration Sensor						
	x	y	z	x,y	x,z	y,z	x,y,z
Accuracy(%)	95.9	74.9	96.3	94.7	97.5	96.1	97.3

Sensor Data for SNN2

	Battery Current	Battery Power	Battery Voltage × Duty of Cutting Motor
Accuracy(%)	96.0	96.6	96.8

Figure 10.
Fusion of sensors for SNN.

features are typically extracted from input signal data during pre-processing. The obtained features are used as input data to a machine learning algorithm. Therefore, for each sensor data, seven neurons are needed in the input layer of SNN shown in **Figure 6**. In the experiments, these features are obtained from a time frame obtained every 3.2 s over 2.3-h measurement data.

The details of collected data are shown in **Table 6**. In the experiments, a subset of time frames from each field data is used for training the SNN, and the obtained model is tested on the remaining test data. The number of time frames used for testing is approximately 670 in each group. The remaining time frames are used for training, as shown in **Table 7**.

5.2.5 Shallow neural network construction

Two SNNs are constructed, and they are cascade connected. The first SNN1 estimates whether the ground is with or without lawn grasses using the horizontal or vertical acceleration sensor. These sensors are expected to measure the acceleration changes caused by the surface of the ground. There are seven feature types, as mentioned in 5.2.4, and 14 neurons in the input layer of SNN1. The second SNN2 estimates the lawn grass lengths using the product of the battery voltage and duty ratio of the motor control signal. This product value tends to vary according to the loads given to the cutting motor. The number of neurons in the input layer of SNN2 is seven. These networks are constructed using "Statistics and Machine Learning Toolbox" in MATLAB [22]. The specifications of the two SNNs are summarized in **Table 8**.

5.2.6 Evaluation results

In the following, two SNNs are first trained, and next, they are used to estimate lawn grass lengths or ground conditions. Their results are compared with those of the RF algorithm. The evaluations are repeated ten times, and their averages are

Groups	Number of time frames
Long Lawn Grasses	2,356
Short Lawn Grasses	1,574
Not Lawn Grasses	2,470
Total	6,400

Table 6.
Specifications of measurement data for evaluating SNN.

Groups	Number of time frames	
	For training	For testing
Long Lawn Grasses	1,686	670
Short Lawn Grasses	904	670
Not Lawn Grasses	1,801	669
Total	4,391	2,009

Table 7.
Number of time frames for training and testing SNN.

	Number of neurons			Used sensors
	Input	Hidden	Output	
SNN1 for Ground Condition	14	10	2	Horizontal or Vertical Acceleration Sensor
SNN2 for Long/Short Lawn Grasses	7	10	2	Battery Voltage & Control Signal's Duty

Table 8.
Configurations of two SNNs.

	Estimation	SNN	RF	Diff.
Precision	Long Lawn Grasses	92.3	94.1	-1.8
	Short Lawn Grasses	96.2	92.8	+3.4
	Not Lawn Grasses	95.8	95.5	+0.3
Recall	Long Lawn Grasses	97.6	94.1	+3.5
	Short Lawn Grasses	90.9	92.4	-1.5
	Not Lawn Grasses	96.3	95.9	+0.4
F-Measure	Long Lawn Grasses	94.9	94.1	+0.8
	Short Lawn Grasses	93.5	92.6	+0.9
	Not Lawn Grasses	96.0	94.1	+0.3
Accuracy		94.8	94.1	+0.7

Table 9.
Evaluation of SNN.

used because the set of training data is randomly selected from the set of time frames, as shown in **Table 7**.

The evaluation results are shown in **Table 9**. The evaluation results corresponding to the evaluation measurements defined in 5.1.3 are shown in this table. The accuracy of the SNN outperforms that of the RF algorithm on average, and the differences in evaluation criteria of each estimation are shown in the "**Diff**" column in **Table 9**. Except for the Precision of "Long Lawn Grasses" and the Recall of "Short Lawn Grasses," the differences are positive. However, the differences are not as large in all estimations, and the required estimation time is negligible.

6. Evaluations of SNN against sensor data stream

As stated in 5.2.6, the SNNs outperforms the RF algorithm, and the estimation system based on the SNNs is implemented on a Raspberry Pi assuming an actual ECU. The processes Lawn Grass Length/Not Lawn Grass Estimator are shown in **Figure 11**. The sensor data streams are sent in a serial format and are received and saved in the memory of the Raspberry Pi. When the required data size is reached, a set of data is preprocessed. Seven features mentioned in 5.2.4 are extracted according to the sensors, including Horizontal/Vertical Acceleration and Battery Voltage times Duty of Cutting Motor, as shown in **Figure 10**. Then, the Lawn Length/Not Lawn Grass Estimator based on the SNNs estimates that a target area is with Long Lawn Grasses, Short Lawn Grasses, or without Lawn Grasses. Finally, the estimation result is sent to the motor controllers, as shown in **Figure 2(b)**.

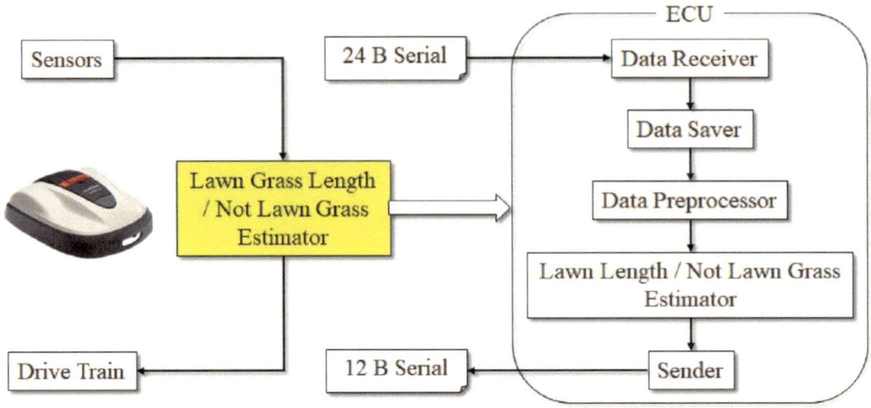

Figure 11.
Implementation of estimator on ECU.

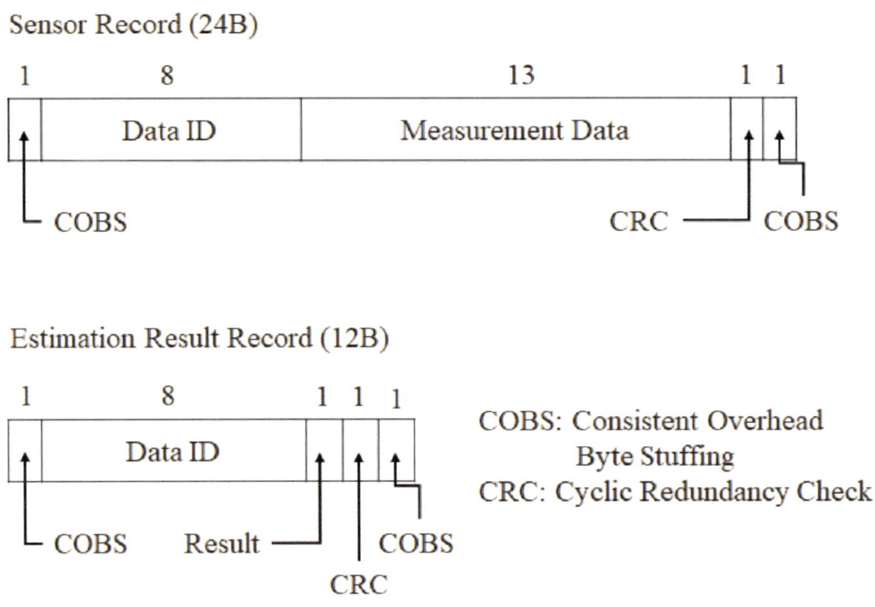

Figure 12.
Frame Design of Sensor and Estimation Result Records.

Figure 12 shows the frames of sensor records and estimation result record. The sensor record has 24 bytes, and it is framed using COBS (Consistent Overhead Byte Stuffing) [23]. This frame has 8-bit Data ID, 13-bit Measurement Data (floating point value), and a 1-bit CRC (cyclic redundancy check) [24] for error detection. The frame of estimation result record consists of COBS, CRC, Data ID, and the estimation result, which is a 1-byte integer (1, 2, or 3) representing the three kinds of estimations.

The sensor data records are obtained every 0.1 s. The 32 records, that is, the 3.2-s measurement data are handled in the SNN training, as shown in **Figure 13(a)**. The robo-mower travels with a speed of 0.55 m/s on average [8], and it moves 1.76 m before the 3.2-s sensor data are obtained. Furthermore, if the total processing time for data I/O, motor control, and wheel driving are assumed to be 2 s, the robo-mower will travel 2.86 m if the estimation waits for a 3.2-s data stream. This distance is too large when the robo-mover reaches the boundary between areas with and without lawn grasses. Therefore, as shown in **Figure 13(b)**, the sensor data stream should be treated

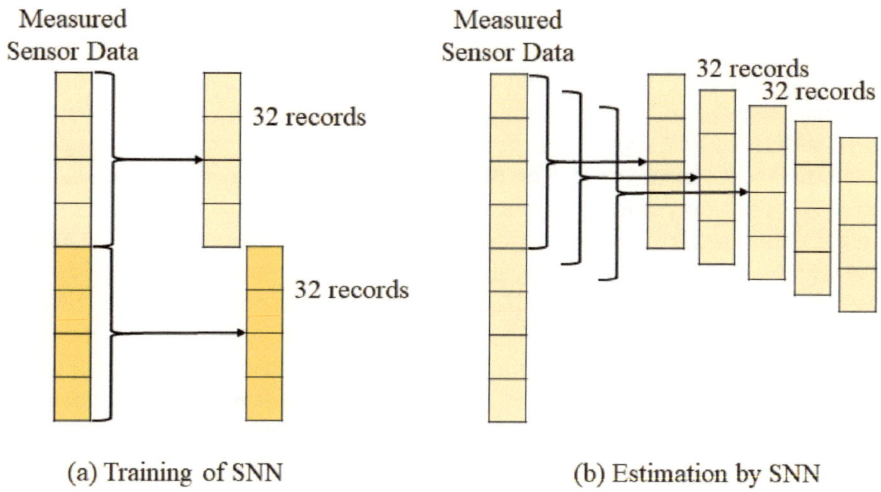

Figure 13.
Pipeline processing of Lawn length estimation.

	Estimation	Pipeline	Non-Pipeline	Diff.
Precision	Long Lawn Grasses	98.9	97.2	+1.7
	Short Lawn Grasses	88.7	90.8	-2.1
	Not Lawn Grasses	94.7	97.4	-2.7
Recall	Long Lawn Grasses	97.6	92.7	+4.9
	Short Lawn Grasses	95.0	96.7	-1.7
	Not Lawn Grasses	89.3	95.5	-6.2
F-Measure	Long Lawn Grasses	98.3	94.9	+3.4
	Short Lawn Grasses	91.7	92.8	-1.1
	Not Lawn Grasses	91.9	95.4	-3.5
Accuracy		94.0	95.0	-1.0

Table 10.
Accuracies between pipeline and non-pipeline processes.

as a pipeline. This means that the estimation is performed every 2.1 s (i.e., 0.1 + 2), resulting in 1.16-m traveling, which is not a significant issue in the real world. The other cause for the estimation delay is the acceleration of other related processes.

Another problem to consider when using the robo-mower in the real world is estimation accuracy. This is because the SNNs are trained with each 3.2-s data frame of sensor data, as shown in **Figure 13(a)**, ensuring that there are no overlaps in the consecutive dataset. The data streams in the pipeline processing have 3.1-s overlap because they are obtained every 0.1 s. The accuracy comparison result is shown in **Table 10**, and the decrement is 1.0 points. Therefore, the estimation accuracy in the pipeline process is not a major issue.

7. Conclusions

The workload estimation methods for autonomous driving of work vehicles are proposed and evaluated. A commercial electric robo-mower is used for the experiments. Specifically, the task to recognize ground conditions, including grounds with long lawn grasses, short lawn grasses, or no lawn grasses, is handled by analyzing data obtained from sensors attached to the robo-mower. Two AI-based algorithms, namely, an RF algorithm and a SNN are proposed. A sensor fusion problem is defined and solved to determine the best combination of sensor data from ten different sensor types. The RF algorithm consisting of 1,000 decision trees and the SNN with only one hidden layer are implemented and evaluated on observation data obtained from various grass cutting field experiments. The RF algorithm achieves 92.3% correct estimation ratio on sensor fusion data in several experiments, while the SNN achieves 95.0%. Furthermore, the accuracy of the SNN is 94.0% in experiments where sensing data are continuously collected as a data stream in real time while the robo-mower is operating. Presently, the proposed estimation system is being developed by integrating two motor control systems into a robo-mower, one for grass cutting and the other for the robot's mobility.

Acknowledgements

The authors express their sincere appreciation to the engineers at Life Creation Center, Honda R&D Co., Ltd., for their technical and financial supports.

Author details

Yoichi Shiraishi*, Haohao Zhang and Kazuhiro Motegi
Graduate School of Mechanical Science and Technology, Gunma University, Ohta, Gunma, Japan

*Address all correspondence to: yoichi.siraisi@gunma-u.ac.jp

IntechOpen

© 2021 The Author(s). Licensee IntechOpen. This chapter is distributed under the terms of the Creative Commons Attribution License (http://creativecommons.org/licenses/by/3.0), which permits unrestricted use, distribution, and reproduction in any medium, provided the original work is properly cited.

References

[1] Zushida K, Zhang H, Shimamura H, Motegi K, Shiraishi Y: Application and Analysis of Random Forest Algorithm for estimating Lawn Grass Lengths in Robotic Lawn Mower. International Journal of Mechanical Engineering and Applications. 2021;9:1:6–14. DOI: 10.11648/j.ijmea.20210901.12

[2] Kojima M, Shimamura H, Motegi K, Shiraishi Y: Comparison of Shallow Neural Network with Random Forest Algorithm in Estimating Lawn Grass Lengths for Robotic Lawn Mowers. In: Proceedings of the SICE Annual Conference 2020; 23–26 September 2020; Thailand; 2020. p. 1634–1639

[3] Ootake K, Shimamura H, Motegi K, Shiraishi Y: A Travelling Simulator for Robotic Lawn Mowers based on Particle Filter Approach. In: Proceedings of the SICE Annual Conference 2020; 23–26 September 2020; Thailand; 2020. p. 1640–1645

[4] Aoki A, Shimamura H, Nakata M, Motegi K, Shiraish Y: Stuck State Avoidance in Work Vehicle's Control with Hybrid Twin Framework. Universal Journal of Control and Automation. 2020;8:2:23–31. DOI: 10.13189/ujca.2020.080201.

[5] Zhang H, Watanabe K, Motegi K. Shiraishi Y: ROS Based Framework for Autonomous Driving of AGVs. In: Proceedings of ICMEMIS; December 2019; 4–6 December 2019; Japan; 2019. IPS6–04

[6] Marc-Oliver S, Max L, Agathe K, Florian K, Breitner, M H: Autonomous Unmanned Ground Vehicles for Urban Logistics: Optimization of Last Mile Delivery Operations. In: Proceedings of the 52nd Hawaii International Conference on System Sciences; 2019;p.1538–1547

[7] Alatise M B, Hancke G P: A Review on Challenges of Autonomous Mobile Robot and Sensor Fusion Methods. IEEE Access. 2020;8:39830–39846. DOI: 10.1109/ACCESS.2020.2975643

[8] Kawakami T, Kobayashi H, Yamamura M, Maruyama S: Development of Robotic Lawnmower Miimo. Honda R&D Technical Review. 2013;25:2:54–59. September 2013 Available from: https://www.hondarandd.jp/point.php?pid=968&lang=en.

[9] Khaleghi B, Khamis A, Karray F O, Razavi S N: Multisensor data fusion: A review of the state-of-the-art. Information Fusion, 2013;14: 28–44

[10] Alam F, Mehmood R, Katib I, Albogam N N, Alebeshr A: Data Fusion and IoT for Smart Ubiquitous Environments: A Survey. Special Section on Trends and Advances for Ambient Intelligence with Internet of Things (IoT) Systems. 2017;5:9533–9554

[11] Takami G, Tokuoka M, Goto H, Nozaka Y: Machine Learning Applied to Sensor Data Analysis. Yokogawa Technical Report English Edition. 2016; 59:1:27–30

[12] Alonso S, Perez D, Moran A, Fuertes J J, Diaz I, Dominguez M: A Deep Learning Approach for Fusing Sensor Data from Screw Compressors. Sensors 2019. 2019;19:13:2868 available from: https://doi.org/10.3390/s19132868–28

[13] Li C, Sanchez R V, Zurita G, Corrada M, Cabrera D: Fault Diagnosis for Rotating Machinery Using Vibration Measurement Deep Statistical Feature Learning. Sensors. 2016;16:895. doi: 10.3390/s16060895

[14] Chinesta F, Cueto E, Abisset-Chavanne E, Virtual L D J: Digital and Hybrid Twins: A New Paradigm in Data-Based Engineering and Engineered Data. Archives of Computational

Methods in Engineering. 2018; DOI: 10.1007/s11831-018-9301-4

[15] El Saddik A: Digital Twins: The Convergence of Multimedia Technologies. IEEE Multimedia. 2018; 25:2:87–92

[16] MPU-9250. available from: https:// www.invensense.com/products/motion-tracking/9-axis/mpu-9250/

[17] Breiman L: Random Forests Machine Learning. 2001;45:5–32

[18] Zhou Z, Feng J: Deep Forest: Towards An Alternative to Deep Neural Networks. In: Proceedings of the Twenty-Sixth International Joint Conference on Artificial Intelligence. 2017;p.3553–3559 doi:10.24963/ijcai.2017/497.

[19] Oka D, Hiroshan Lakmal Balage D, Motegi K, Kobayashi Y, Shiraishi Y: A Combination of Support Vector Machine and Heuristics in On-line Non-Destructive Inspection System. In: Proceedings of the International Conference on Machine Learning and Machine Intelligence (MLMI). September 2018; Vietnam; 2018. CM0007 archived in IEEE Xplore Digital Library

[20] H. Witten I, Eibe Frank E, A. Hall M: Data Mining: Practical Machine Learning Tools and Techniques: Morgan Kaufmann; 2011

[21] TreeBagger, Math Works. available from: https://jp.mathworks.com/ help/stats/treebagger.html?lang=en.

[22] Statistics and Machine Learning Toolbox, MathWorks. available from: https://www.mathworks.com/products/statistics.html#machine-learning

[23] Consistent Overhead Byte Stuffing. available from: https://en.wikipedia.org/wiki/Consistent_Overhead_Byte_Stuffing

[24] Cyclic redundancy check. available from: https://en.wikipedia.org/wiki/Cyclic_redundancy_check

Chapter 9

A Distributed Approach for Autonomous Cooperative Transportation

Amar Nath and Rajdeep Niyogi

Abstract

Autonomous mobile robots have now emerged as a means of transportation in several applications, such as warehouse, factory, space, and deep-sea where direct human intervention is impossible or impractical. Since explicit communication provides a better and reliable way of multi-robot coordination compared to implicit communication, so it is preferred in critical missions, such as search and rescue, where efficient and continuous coordination between robots is required. Cooperative object transportation is needed when the object is either heavy or too large or needs extra care to handle (e.g., shifting a glass table) or has a complex shape, which makes it difficult for a single robot to transport. All group members need no participation in the physical act of transport; cooperation can still be achieved when some robots transport the object, and others are involved in, say, coordination and navigation along the desired trajectory and/or clear obstacles along the path. A distributed approach for autonomous cooperative transportation in a dynamic multi-robot environment is discussed.

Keywords: cooperative transportation, distributed algorithm, dynamic environment, multi-agent coordination and cooperation

1. Introduction

Autonomous mobile robots are now used in several applications, such as warehouse, factory, space, and deep-sea, that may be inaccessible for humans. The main concern is to find an effective coordination mechanism among autonomous agents to perform tasks in order to achieve high quality overall performance. Although MAS research has received substantial attention, multi-robot coordination remains a challenging problem since the overall performance of the system is directly affected by the quality of coordination and control among the robots while executing cooperative tasks. Coordination in a multi-robot system can be achieved either by explicit or by implicit communication. Since explicit communication provides a better and reliable way of multi-robot coordination compared to implicit communication, so it is preferred in critical missions, such as search and rescue, where efficient and continuous coordination between robots is required.

A collaborative task cannot be executed by any single agent. It requires multiple agents at the task's location. Execution of such tasks is quite challenging in a dynamic environment, as the time and location of a task arrival, required skills, and the number of robots required for its execution may not be known a priori. This necessitates the design of a distributed algorithm for collaborative task execution

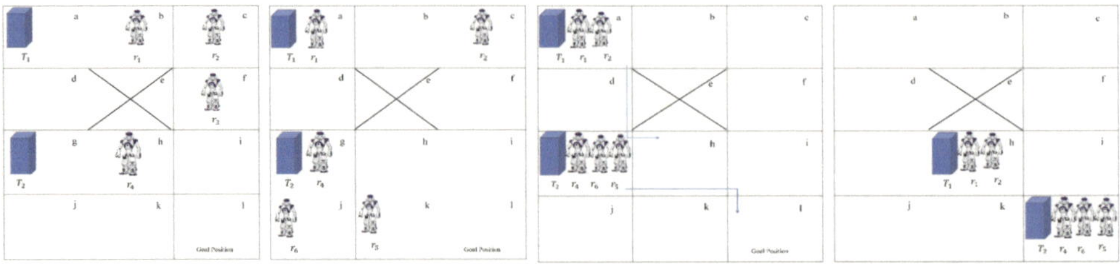

Figure 1.
Snapshots of a dynamic environment.

via runtime team/coalition formation. To form a team with a lack of global knowledge, the robots need to communicate with each other to acquire relevant information. Here, a distributed approach for collaborative task execution in a dynamic environment is discussed. We illustrate the applicability of the approach with urban search and rescue (USAR) domain and evaluate its performance with extensive experiments using ARGoS, a realistic multi-robot simulator.

We now illustrate an example scenario of the problem considered as shown in **Figure 1**. The environment, a grid world of size 4×3, consists of 12 locations marked as $a, b, \dots, l$. Robots can move to its adjacent location. Boxes arrive at the locations at different points in time. The task is to move a box from its arrival location to the goal location, which is marked on the box.

The snapshots of the environment at different instants of time is shown in **Figure 1**. At time t_1, two tasks τ_1 and τ_2 arrive at locations a and g respectively, 4 robots are present at the locations b, c, f and h. The robots r_1 and r_4 detect the tasks τ_1 and τ_2 respectively. At time t_2, r_1 and r_4 move to locations a and g to attend the tasks. Now, r_1 and r_4 determine the team sizes to be 2 and 3, and the goal locations to be h and l for the tasks τ_1 and τ_2 respectively. At this time, r_3 exits and r_5 and r_6 enter at locations k and j respectively.

At t_2, r_1 and r_4 do not know the states and locations of other robots present in the environment, and thus with this insufficient information they cannot form their respective teams. Thus, in order to form their teams, they invoke the algorithm given in Section 4. At t_3, r_1 and r_4 both form their teams successfully and the members reach the locations of the tasks as shown in the Figure. Finally, at time t_4, execution of the tasks are completed and the team members for τ_1 and τ_2 reach their respective goal locations.

2. Related work

In the literature, several approaches have been suggested for solving the problem of cooperative object transportation [1–4]. The work [1] is considered as the pioneering work, targeting a cooperative transport task by a homogeneous group of simple robots that can only push an object. The authors [1] demonstrate that coordinated effort is not possible without explicit communication.

The work [2] proposed direct (explicit) communication to improve the coordination of a homogeneous group of two six-legged robots required to transport a rectangular box towards a target cooperatively. The work [3] considered the problem of cooperative box pushing where the roles of the members are pre-defined; specifically one robot acts as a watcher and the others act as pusher. However, we consider a more complex scenario of cooperative object transportation scenario, where the role of each robot is not fixed in advance, rather decided at runtime. In [4], the robots are designed to push the object across the portion of its surface, where it occludes the direct line of sight to the goal. This simple behavior results in transporting the object towards the goal without using any form of direct communication.

The problem of cooperative transportation, considered in this paper, involves team formation of heterogeneous robots and gathering the robots to the location of the object to be transported. The number of robots required to transport the object is not known a priori and it is decided at runtime. For the same task, the team size is determined by the state of the environment. So, at some point in time an object may be transported by two robots while at some other moment in time three or more robots are required. Few works related to coalition formation strategies are discussed below.

Auction-based approaches for team formation (task allocation) are suggested in [3, 5]. A bidder agent has some resources (e.g., data center, CPU) [5], who may bid for multiple auctioneers concurrently. However, when we move to physical agents, a robot cannot be a member of multiple coalitions at any point of time simply because the tasks may be at different locations, and a robot cannot be at two different locations at the same time, even though a robot may have the capability to perform multiple tasks at a time.

In our work, a non-initiator robot (bidder) will not express its willingness to multiple initiators (auctioneers) concurrently; when more than one request message arrives, the robot stores the requests in its local queue. Having one or more resources specified in the auction is a sufficient condition for an agent to make a bid [5]. Having the required skills for a task is a necessary but not a sufficient condition for a robot to express its willingness to be part of a team, in our work. A robot's behavior, in our work, is determined by its current state, whereas in [3, 5] states need not be taken into consideration.

In [6], the authors describe a framework for dynamic heterogeneous team formation for robotic urban search and rescue. The task discovery is made by a member of a team and it is sent to the team coordinator for assignment. The team coordinator performs the task assignment ensuring the task will be carried out by a robot with the necessary capabilities. However, in a distributed system, no robot knows the states, locations, and skills of other robots. Thus, the robots should communicate among themselves to acquire relevant information for task execution without the intervention of any central authority. This necessitates the design of a distributed algorithm for task execution in such a dynamic environment.

In our approach, unlike [6], every robot has a similar level of priority, and each of them can perform the task management activities, i.e., searching, team/coalition formation by acquiring the information from the robots available in the environment at that moment in time. In this paper, the arrival time and location of a task are not known a priori; hence, task searching and coalition formation activities are performed by a robot at runtime.

3. Problem formalization

A formal framework of a dynamic environment and some related concepts are presented below.

Definition 3.1. (Dynamic environment) A global view (snapshot) of an environment $\mathcal{E}$, with a set of locations L, taken at time t, is given by a 3-tuple $\mathcal{E}^t = \langle \mathcal{R}^t, \mathcal{T}^t, f \rangle$, where $\mathcal{R}^t$ is the set of robots present in the environment at time t, and $\mathcal{T}^t$ is the set of tasks that arrive in the environment at time $t, f : \mathcal{R}^t \times \mathbb{N} \mapsto L$, is a function that gives the location of a robot at a discrete instant of time represented by the set of natural numbers $\mathbb{N}$.

A robot has a set of skills ψ (eg., gripper, camera), and at any instant of time it may be in any state from the set of states $\mathcal{S} = \{Idle, Ready, Promise, Busy\}$. A robot can enter the environment $\mathcal{E}$ at any time, but can leave only if its state is *Idle*. When a robot attends a task, it can determine the information required to begin team formation, from the task specification, which is given below.

Definition 3.2. (Task (cooperative transportation)) A task τ is specified by a 5-tuple $\tau = \langle \nu, l, t, k, \Psi \rangle$ where ν is the name of a task (e.g., move (carry) box B to location l', lift desk D), $l \in L$ is the location where the task arrived, t is the time at which the task arrived, $k > 1$ is the number of robots required to execute the task, and Ψ is the set of skills required to execute the task.

Definition 3.3. (Condition for single task execution) A task $\tau = \langle \nu, l, t, k, \Psi \rangle$ can be executed, if there exists a set $\mathcal{R}$ of k available robots, such that for all $r \in \mathcal{R}$, $\psi_r \supseteq \Psi$ at some time $t' > t$, and for all $r \in \mathcal{R}$, location of r, $loc_r = l$ at some time $t'' > t'$.

The first condition in the *if* is for team formation, and the second condition is for ensuring that all the team members converge to the location of the task.

Definition 3.4. (Condition for multiple task execution) The tasks $\tau_1 = \langle \nu_1, l_1, t, k_1, \Psi_1 \rangle$ and $\tau_2 = \langle \nu_2, l_2, t, k_2, \Psi_2 \rangle$ can be executed if the following conditions hold:

1. there exists a set $\mathcal{R}_1$ of k_1 available robots, such that for all $r \in \mathcal{R}_1$, $\psi_r \supseteq \Psi_1$ at some time $t_1' > t$, and for all $r \in \mathcal{R}_1$, $loc_r = l_1$ at some time $t_1'' > t_{1'}$.

2. there exists a set $\mathcal{R}_2$ of k_2 available robots, such that for all $r \in \mathcal{R}_2$, $\psi_r \supseteq \Psi_2$ at some time $t_2' > t$, and for all $r \in \mathcal{R}_2$, $loc_r = l_2$ at some time $t_2'' > t_2'$.

3. $\mathcal{R}_1 \cap \mathcal{R}_2 = \emptyset$.

Definition 3.5. (Utility of a team for task execution) Let $\Gamma = \{x_1, \ldots, x_k\}$ be a team that can execute a task $\tau = \langle \nu, l, t, k, \Psi \rangle$ where each member of the team was located at loc_{x_i}. The utility of a team Γ for executing τ is $\mathcal{U}_{\langle \Gamma, \tau \rangle} = -cost_{\langle \Gamma, \tau \rangle}$, where $cost_{\langle \Gamma, \tau \rangle} = \sum_{x_i \in \Gamma} \mu_{\langle x_i, \tau \rangle}$ and $\mu_{\langle x_i, \tau \rangle} = \mathbf{p}(x_i, \tau) \times \frac{1}{\alpha_{x_i}} + \mathbf{d}(loc_{x_i}, l) \times \beta_{x_i}$.

where $\alpha_{x_i}, \beta_{x_i} \in (0, 1]$ denote remaining battery coefficient and battery consumption rate respectively of (a robot) x_i, $\mathbf{p}(x_i, \tau)$ is the price of x_i for τ, $\mathbf{d}(l_1, l_2)$ is the distance covered when moving from l_1 to l_2.

A robot with higher α value ensures that it will not fail due to its more remaining battery backup. A robot with lower β value ensures that it will last for a longer period of time.

4. Distributed algorithm for cooperative transportation

Following assumptions are made for the study. Multiple robots are required for any task execution. A robot can execute at most one task at a time. Each robot has a unique identifier (id). A wireless network that is lossless, message delay is finite, data is not corrupted during transmission is considered. Messages are delivered in a FIFO manner.

Informal description of the algorithm is given below. Let a robot i attend a task $\tau = \langle \nu, l, t, k, \Psi \rangle$ where $\psi_i \supseteq \tau.\Psi$. To execute the task (cooperative transportation), initiator communicates with other robots in order to form a runtime team. Here, the i is named as an initiator, and the other robots as non-initiators.

After task detection, i broadcasts a *Request* message to know the current state of the other robots present in the environment at that moment in time and waits for some time, say Δ. The broadcast messages are delivered only to those robots who are present in the range. Now, on receipt of *Request* message, a non-initiator j takes the necessary actions. A non-initiator who has the desired skill will send a *Willing* and an *Engaged* message if its state is other then *Idle*.

The counter c of initiator is increased on receipt of *Willing* message. The parameter $(c \geq k - 1)$ is checked after Δ time has elapsed. Now, if the value of $(c \geq k - 1)$ is true then team formation that has maximum utility is possible and sends *Confirm* message to the members of the team and sends a *Not-Required* message to $(c - (k - 1))$ robots, if any. However, if the value of the condition $(c \geq k - 1)$ is false, i sends a *Not-Required* message to all c robots who expressed their willingness to help. Also, i changes its state from state *Ready* to *Idle* The algorithm has the non-blocking property since a timer is used. If there was no timer, an initiator would have waited indefinitely and thereby forcing some non-initiators to wait indefinitely as well; thus the system would be blocked.

The *receive* function of a robot is given in Algorithm 2. The agents take the action based on the current state that may be *Idle* (line 17–21), *Promise* (line 22–39), *Busy* (line 12–16 and 41–45), and *Ready* (line 1–11). Within a state, the type of message is checked and appropriate actions are taken. For example, if an agent receives a *Request* message in *Idle*, the identifier of the sender is enqueued, and *flag* is set to true; if it has appropriate skills then it sends the *Willing* message to the sender (initiator) and *flag* is set to false.

The behavior of the agent is captured with communicating automata (CA) [7] as shown in **Figures 2** and **3**. Moreover, this communicating automata is helpful in understanding and designing the algorithm.

Transitions in CA are very general form $\chi : \gamma$, where χ can either be an input a (send message $!m$, receive message $?m$), or a state condition g, or (a,g), and γ can

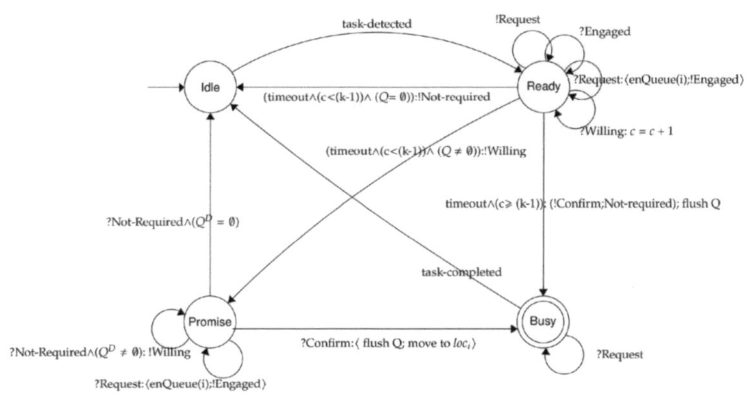

Figure 2.
Finite state machine for an initiator agent.

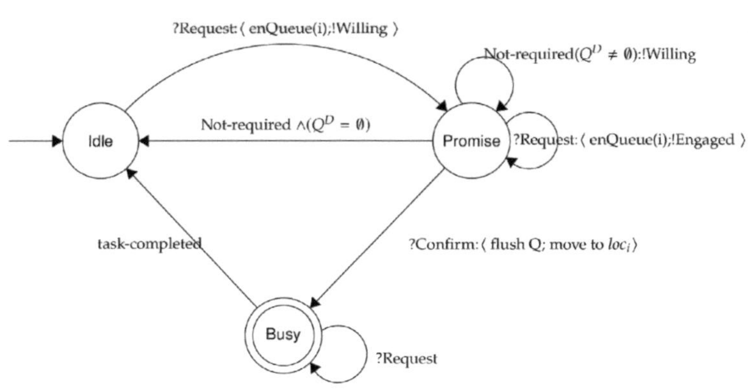

Figure 3.
Finite state machine for a non-initiator agent.

either be a sequence of actions *seq*, or a sequence of actions that is to be performed atomically $\langle seq \rangle$, or empty. Similarly, semantics are defined.

4.1 Analysis of the algorithm

4.1.1 Message complexity

Let there be I initiators at some instant of time, say t. Each initiator broadcasts a *Request* message, which is sent to $(N-1)$ robots, where N is the total number of robots present at time t. So, the total number of such messages would be $(N-1) \cdot I$ which is $O(N \cdot I)$. The total number of replies obtained from non-initiators would be at most $(N-1) \cdot I$ which is $O(N \cdot I)$. An initiator sends c number of *Confirm* and *Not-Required* messages, which is $O(N)$. Thus total messages send by all the initiators would be $O(N \cdot I)$. Thus the total number of messages would be the sum of these messages, and this becomes $O(N \cdot I) + O(N \cdot I) + O(N \cdot I)$, which is $O(N \cdot I)$. When the number of initiators is relatively small compared to the total number of robots present at time t, the message complexity would be $O(N)$.

4.1.2 Handling multiple initiators

Let us consider the snapshot of the environment at t_2 in **Figure 1**, where r_1, r_4 invoke the *send* function (Algorithm 1) simultaneously; r_1, r_4 need one, two other robots respectively. The initiators r_1, r_4 broadcast *Request* messages corresponding to their respective tasks. Let all the other robots be in *Idle* state initially and they can satisfy the requirements of both the tasks. Eventually r_2 becomes part of the team with r_1 because it received the *Request* message from r_1 before it received the *Request* message from r_4. Similarly, eventually r_5 and r_6 become part of the team with r_4.

Algorithm 1: (Send function of i)

```
Send function of initiator i for a task τ = ⟨v, l, t, k, Ψ⟩ :
state := Ready;
c := 0;
broadcast Request_i;
set the timer and wait till timer is OFF;
if c ≥ (k − 1) then
    for each possible team, calculate utility as per Defn. 3.5;
    Select the team that has highest utility, say Γ;
    send Confirm_i to the robots of the team Γ;
    send Not-Required_i to the other c − (k − 1) robots;
    ⟨state := Busy; make Q_i empty⟩;
    initiate execution of task τ when all members of Γ arrive at the location l;
else
    send Not-Required_i to c robots;
    if Q_i = ∅ then
        state := Idle;
    else
        state := Promise;
        send Willing(i) to the front element of Q_i;
    end
end
if a Willing message is received in a state ≠ Ready from a robot j then
    send Not-Required_i to j;
end
Send function of Non-initiator j for a task τ = ⟨v, l, t, k, Ψ⟩ :
case Q_j ≠ ∅ and flag = true do
    send Willing_j to the front element of Q_j;
    flag := false;
end
case Q_j ≠ ∅ and flag′ = true do
    send Engaged_j to the rear element of Q_j;
    flag′ := false;
end
```

DOI: http://dx.doi.org/10.5772/intechopen.98270

The complete execution trace of algorithms 1,2 is shown in **Figure 4** using message sequence chart (MSC).

Algorithm 2: (Receive function of j)

```
Receive function of Initiator j for a task τ = ⟨v, l, t, k, Ψ⟩:
case state = Ready do
    case msg = Request do
        ⟨enqueue(Q_j, i); flag' := true⟩
    end
    case msg = Willing do
        c := c + 1;
    end
    case msg = Engaged do
        skip;
    end
end
case state = Busy do
    case msg = Request do
        skip;
    end
end

Receive function of Non-initiator j for a task τ = ⟨v, l, t, k, Ψ⟩ :
case state = Idle and ψ_j ⊇ τ.Ψ do
    case msg = Request do
        ⟨state := Promise; enqueue(Q_j, i) if i is not present in Q_j; flag := true⟩
    end
end
case state = Idle and ¬(ψ_j ⊇ τ.Ψ) do
    skip;
end
case state = Promise and ψ_j ⊇ τ.Ψ do
    case msg = Request do
        ⟨enqueue(Q_j, i) if i is not present in Q_j; flag' := true⟩
    end
    case msg = Confirm do
        ⟨state := Busy; make Q_j empty; move to loc_i⟩
    end
    case msg = Not-Required do
        dequeue(Q_j);
        if Q_j = ∅ then
            state := Idle;
        else
            flag := true;
        end
    end
end
case state = Busy do
    case msg = Request do
        skip;
    end
end
```

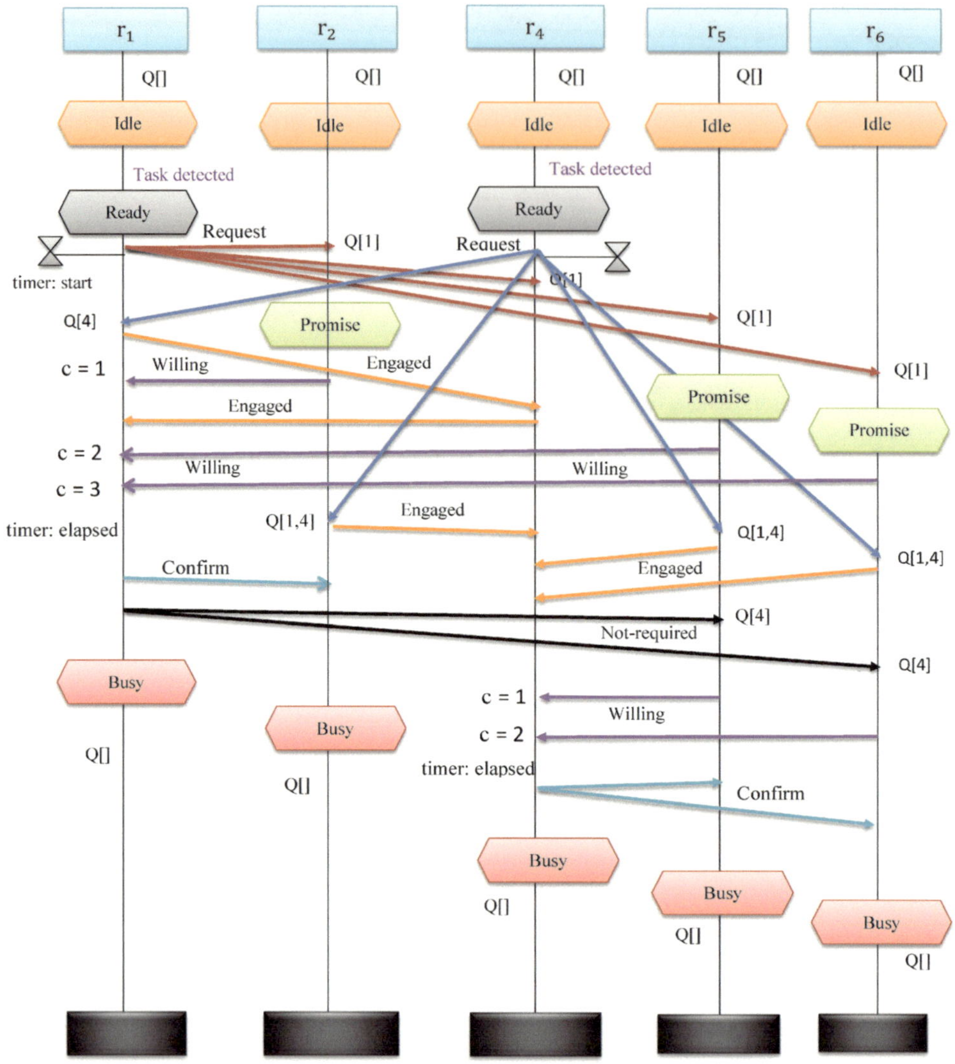

Figure 4.
Execution trace of the algorithms for multiple initiators.

5. Implementation in ARGoS

We consider a road clearance scenario to illustrate the proposed distributed algorithm (Section 4), where a road may be blocked by several obstacles. A team of robots should jointly move each obstacle to one side of the road. The algorithm is implemented using ARGoS (Autonomous Robots Go Swarming) [8], a multirobot simulator using the 3.0.0-beta47 version on Intel⊕ Core™ i5 Processor, 4-GB of RAM and macOS Sierra operating system. The code run in ARGoS can be directly deployed on a real robot system.

An example scenario is shown in **Figure 5**, where the shaded portion in gray is the road (10 m × 5 m), obstacles are simulated by green movable cylinders of radius 0.2 m with a blue light on top. The robots are shown in blue. The overall process of removing an obstacle from the road is shown in **Figure 5**. The robots in ARGoS use the inbuilt range and bearing sensor (*rab*) to communicate among themselves.

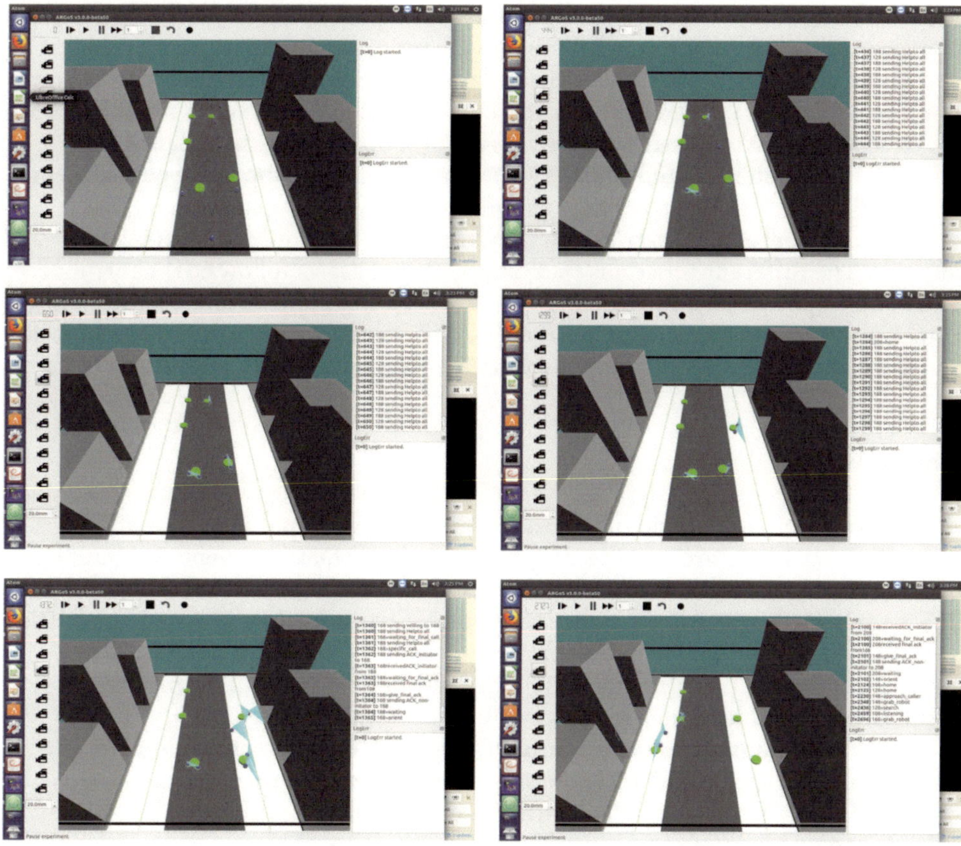

Figure 5.
Illustration of multiple task execution in ARGoS.

The broadcast of the messages to all other robots is done by *rab* actuator. The broadcast of message is done within a certain range and in line of sight. We have used the 3 bytes for message within the range of 15 meters. The message is received by *rab* receiver within in the same network sent by *rab* sensors. Along with sending and receiving the message within range, *rab* sensors do the work of identifying the direction and distance from where the message is being sent. As the *rab* actuator allows the only broadcast, the address of the sender and that of the receiver needs to be specified in every message. Every robot in the simulation has a unique id of size 1 byte. Several sensors and actuators are used to control the movement and positioning of the robots. For example, proximity sensors are used to stay on the road and avoiding collisions with other robots, the omni-directional sensor is used to detect obstacles, gripper actuator is used to grip an obstacle, and turret actuator is used to turn the gripper actuator towards the direction of the obstacle.

In **Figure 5a**, the initial position of the robots and blocks is shown. Three robots detect the three obstacles and they start the formation for the same is shown in **Figure 5b**. We assume that all the obstacles require two robots to move. In **Figure 5c**, two initiator robots are able to form their teams. In **Figure 5d**, it is depicted that robots have reached to the location of obstacles and they are ready to move the obstacles. **Figure 5e**, clearly shows that both the obstacles have been shifted to one side of the road. After dropping the obstacles, the robots again visit the road and search for other obstacles if any. Finally, in **Figure 5f**, the third obstacle is also detected and removed. In this way, all the obstacles are removed from the road.

For the implementation we have written the required functions in Lua (a C-like language). These are: (i) to control the movement of a robot to avoid obstacle or another robot based on proximity sensor data, where the sensor detects an obstacle or another robot, (ii) control speed and velocity, (iii) synchronizing the robots for task execution, (iv) to control the movement of a robot when boundaries are detected using motor-ground sensors, (v) communication among robots based on the line of sight.

The implementation is carried out by writing the required function in Lua language. The different functions that are identified are as follows: (i) control of velocity and speed of the robot, (ii) control the movement of a robot so that obstacles and other robots could be avoided, (iii) synchronizing the robots in order to task execution, and (iv) communication among robots based on the line of sight.

6. Summary

Now, research in the field of robotics is going with a rapid rate. In many applications such as search and rescue, space, and automated warehouse, intelligent robots are being used. With the advancement of artificial intelligence domain, robots are becoming the good choice. A plenty of work has been carried out in the field of single robot. However, this chapter discuss the different aspects of work where multiple robots act on the same object at the same time. This problem becomes tough and different from normal multi-agent problem.

Cooperative transportation is common task in many challenging domains, i.e., rescue, mars and space, and autonomous warehouse etc. In this way the proposed framework becomes very much essential and important in such domains where multiple robots are required to execute a task.

The proposed approach also takes care of multiple task execution simultaneously, i.e., if multiple robots detect multiple different obstacles at the same time, the coalition formation process for each obstacle can be started. Each robot who detects the obstacle, starts the coalition formation, by executing the instance of the algorithms.

Author details

Amar Nath[1,2*†] and Rajdeep Niyogi[2†]

1 Sant Longowal Institute of Engineering and Technology, Punjab, India

2 Indian Institute of Engineering and Technology, Roorkee, India

*Address all correspondence to: amarnath@sliet.ac.in

† These authors contributed equally.

IntechOpen

© 2021 The Author(s). Licensee IntechOpen. This chapter is distributed under the terms of the Creative Commons Attribution License (http://creativecommons.org/licenses/by/3.0), which permits unrestricted use, distribution, and reproduction in any medium, provided the original work is properly cited. (cc) BY

References

[1] Kube, C. R., & Zhang, H. (1993). Collective robotics: From social insects to robots. Adaptive behavior, 2(2), 189-218

[2] Mataric, M. J., Nilsson, M., & Simsarin, K. T. (1995, August). Cooperative multi-robot box-pushing. In Proceedings 1995 IEEE/RSJ International Conference on Intelligent Robots and Systems. Human Robot Interaction and Cooperative Robots, Vol. 3, pp. 556-561

[3] Gerkey, B. P., & Mataric, M. J. (2002). Sold!: Auction methods for multirobot coordination. IEEE transactions on robotics and automation, 18(5), 758-768

[4] Chen, J., Gauci, M., Li, W., Kolling, A., & Gro, R. (2015). Occlusion-based cooperative transport with a swarm of miniature mobile robots. IEEE Transactions on Robotics, 31(2), 307-321

[5] Kong, Y., Zhang, M., & Ye, D. (2016). An auction-based approach for group task allocation in an open network environment. The Computer Journal, 59(3), 403-422

[6] Gunn, T., & Anderson, J. (2015). Dynamic heterogeneous team formation for robotic urban search and rescue. Journal of Computer and System Sciences, 81(3), 553-567

[7] Brard, B., Bidoit, M., Finkel, A., Laroussinie, F., Petit, A., Petrucci, L., & Schnoebelen, P. (2013). Systems and software verification: model-checking techniques and tools. Springer Science & Business Media

[8] Pinciroli, C., Trianni, V., OGrady, R., Pini, G., Brutschy, A., Brambilla, M., Dorigo, M. (2012). ARGoS: a modular, parallel, multi-engine simulator for multi-robot systems. Swarm intelligence, 6(4), 271-295

Chapter 10

Interaction Protocols for Multi-Robot Systems in Industry 4.0

Edi Moreira M. de Araujo, Augusto Loureiro da Costa and Alejandro R.G. Ramirez

Abstract

In this chapter, the main methods of communication among multi-robot systems involved in Machine-to-Machine (M2M) applications, especially with regard the communication, reliability, stability and security among these robots, presenting various concepts through papers already published. A comparative study was carried out between two communication protocols applied in M2M technologies, the Queue Telemetry Transport (MQTT) developed by IBM along with Eurotech and the Constrained Application Protocol (CoAP). A study and survey of the characteristics of each of the protocols was carried out, as well as the method of operation of each of them and how both can be used in applications involving multiple robots. It was concluded that both protocols are considered ideal for use in in applications involving multi-robot systems. However, although the two protocols have been designed for application in environments with limited communication, the MQTT exchange protocol has advantages over CoAP, as a lower ovehead between message exchanges.

Keywords: Machine-to-Machine, protocol, multi-robot system

1. Introduction

The industry, in the last century, has undergone changes in the way it operates, generating innovation and profound social and economic changes. According to [1], is the beginning of a revolution called Industry 4.0. This industrial revolution is based on several concepts, among them, the Cyber-Physical System, Internet of Things (IoT), big data analytics, Machine-toMachine (M2M) and cloud computing. All of these concepts aim to meet the requirements of an advanced manufacturing system, promoting the integration of an entire supply chain.

The authors in [2], Industry 4.0 creates what has been called the smart factory. This factory has a modular structure in which cyber-physical systems monitor physical processes, creating a virtual copy of the physical world, making decentralized decisions using the IoT that has communication with each other and with humans in real time. These smart factories aim to solve several challenges found in large industrial systems, due not only to the increase in the complexity of processes and products, but also to the increase in the varieties of these products, which must be placed on the market due to the reduced life cycle. Thus, there is a need to make production processes more flexible, characterized according to [3] in production systems with distributed units, composed of a conglomerate of autonomous units, which operate in cooperation in an integrated manner. Such units can be industrial automation machines, manipulating robots, mobile robots or microprocessed remote units.

Distributed production systems, often composed of robot machines, are designed with the objective of providing efficiency and rationality in the use of distributed production resources, in order to favor the manufacture of products, in a dynamic and fast way. The production units must be able to respond, in an intelligent and effective manner, to unforeseen disturbances in the external environment, maintaining controlled and continuous production [3]. Considering the need to plan and control systems for these units, complete robotization of productive systems, which in turn need means or protocols of interaction and coordination between them.

Therefore, the justification for proposing this chapter is to deepen the studies on the interaction protocols for existing multi-robot systems and to design a new protocol that can be applied to concepts related to Industry 4.0, taking into account the characteristics of self- organization of robotics structure based on the concept of industrial agents.

This chapter is divided as follows. In Section 2, the Machine-to-Machine (M2M) is presented, with its levels explained. In Section 3, protocols MQTT and CoAP are presented, identifying their main characteristics and limitations. A comparison between the protocols (MQTT and CoAP) will be demonstrated in the Section 4. Section 5 is shown some studies that used MQTT protocol, along with Robot Operating System (ROS) in the context of Insdustry 4.0, in addition to presenting the conclusions of the chapter.

2. Machine-to-machine communications

According to [4] the term M2M Communications, it is the machine to machine communication, which enables the transmission of data across different devices without the need for human intervention.

This communication opens up an immense range of applications that can, among other things, register, process and manipulate the data generated and transmitted by the objects that are interconnected. For example, an application that continuously receives data from a sensor that measures the temperature of an environment and, based on the data obtained, can generate statistics that describe the sensor readings over a period of time and then send an alert via e-mail or Short Message Service (SMS) to one or more individuals if the temperature has reached very high levels, or even publish this information to another device that could use it in another way, among other things.

M2M applications have the potential to become a trend in the development of software in the coming years in view of the various sectors (such as industrial and home automation) that need an automated solution that integrates the devices that are part of their environment. Devices that are part of an M2M network have the ability to at least collect data from a given environment and transmit it to an application through a connection. Eventually, these devices will not be able to transmit this data directly to other equipment, it is necessary to use a gateway to be an intermediary for this transmission.

Thus, the M2M can be defined as a number of technologies that aims to establish communication between devices with the ability to transmit information for a particular application without the interference of a human action.

2.1 M2M architecture

According to [4], the M2M architecture (**Figure 1**) is divided into three domains, Devices, gateway and Network. The components of these domains are described as follows:

- In the domain of devices:

  M2M device: A device that runs one or more M2M applications using M2M service capabilities. The M2M device connect to network domain in the following for two manners, by Direct Connectivity (M2Mdevices connect to the network domain via the access network) and Gateway as a Network Proxy (The M2M device connects to the network domain via an M2M gateway);

- M2M Gateway: object that runs M2M applications, using M2M services and acting as a proxy between M2M devices and the network domain;

  Core network: Its main function is to ensure the functioning of the network with connectivity via IP and other means of connectivity, as well as control functions of network services, interconnection and roaming;

  Access network allows M2M devices and gateways to communicate with the core network. According to [5], examples of M2M include technologies such as IEEE 802.15.1, Bluetooth, personal area network, among others, or local networks such as power line communication with PLC and Wireless M-BUS;

- M2M area Network:

  M2M Network Area: provides the connection between devices and gateways;

  Provide M2M functions that are shared by different applications;

  M2M applications: run the service logic;

  Network Management: brings together all the functions necessary to manage the network core and the access network;

  M2M management roles: Consist of the roles needed to manage service capabilities within the network domain;

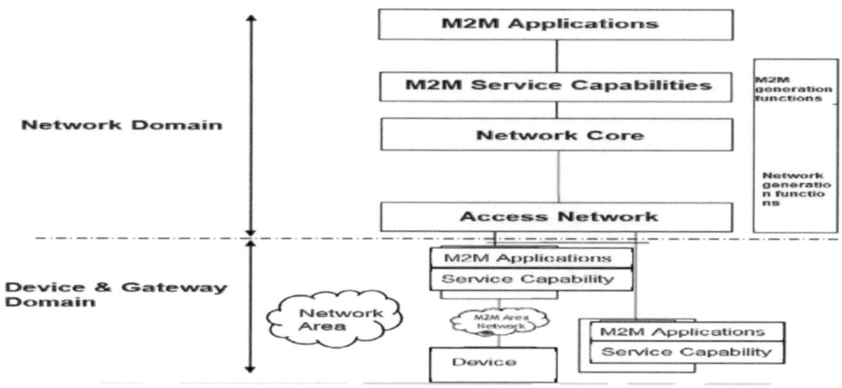

Figure 1.
M2M architecture.

In **Figure 2** a simplified version of the M2M architecture is presented, where important elements of the architecture are shown, in addition to defining the application domain.

2.2 M2M systems categories

The authors in [6] categorize M2M systems into two types, namely dynamic M2M systems and static M2M systems. The main difference between the two is its topology, where in dynamic systems, some nodes (for example, M2M devices and M2M gateways) are moving, that is, the topology is changing over time, resulting in a change in the quality of communication, and dynamic resource allocation. Examples of dynamic M2M systems include: vehicle M2M system, the medical M2M system and the robotic M2M system. In contrast, the topology for static M2M systems remains unchanged for a relatively long time, as an example the M2M power system, domestic M2M system and the industrial M2M system.

3. M2M communication protocols

There are several communication protocols responsible for managing the transfer of data between computers on the internet, among them we can mention some examples such as HTTP (Hypertext Transfer Protocol), FTP (File Transfer Protocol) and SFTP (SSH File Transfer Protocol). When the communication needs to be made between two or more devices (or several applications) connected in a network, the need arises to have a protocol that manages this communication, that is, the exchange of data between the devices in an efficient way considering the characteristics and restrictions imposed by the environment. According to [4], in this scenario, two protocols arise that can be used in restricted environments: Messaging Queue Telemetry Transport (MQTT) and the Constrained Application Protocol (CoAP).

The following will present the MQTT and CoAP protocols, identifying their main characteristics and limitations, in addition to highlighting the best scenarios where each can be applied in the context of Industry 4.0.

3.1 MQTT: message queue telemetry transport

Created by IBM in 1999, MQTT is an open source protocol designed to be simple, lightweight and easy to implement. It is a messaging protocol based on the

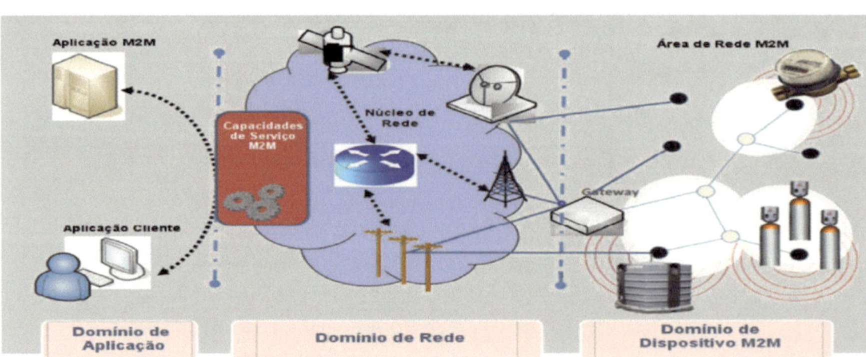

Figure 2.
Simplified version of high-level architecture.

publish/subscribe architecture (**Figure 3**), which has a small transport overhead (fixed byte header of 2 bytes), making MQTT an interesting solution for unreliable networks with limited resources, such as bandwidth and high latency [4]. This protocol is based on a broker (**Figure 4**) using the message pattern *publish/subscribe*, while the server broker acts as a intermediary for messages sent from a device that publishes to subscribing customers, providing a distribution of one-to-many messages decoupled from the use case of the application.

For a client to send a message, it needs to publish it in a topic (called a MQTT broker) (**Figure 4**). If another client wants to receive the content of this message, he will have to subscribe to this same topic. A client can publish or subscribe to multiple topics at the same time, and there may be situations where the publication or subscription on a topic is disputed between different clients, thus having a system that is asynchronous [7].

The PDU (Protocol Data Unit) of the MQTT protocol is encapsulated by the TCP (Transmission Control Protocol) protocol, that is, the MQTT header and data are sent in the TCP data area [8]. In this way, the MQTT protocol messages have a fixed header (**Figure 5**) composed of two bytes, where the first byte contains the field that identifies the type of message, such as also the markers (DUP, QoS level and RETAIN). There is a version of MQTT, called MQTT-SN (MQTT Sensor Network), where PDU is encapsulated by the UDP protocol, which, in turn, is encapsulated by the IP or the 6LowPAN protocol. One of the main differences between two standards, in addition to the network layer they focus on, is the simplification of

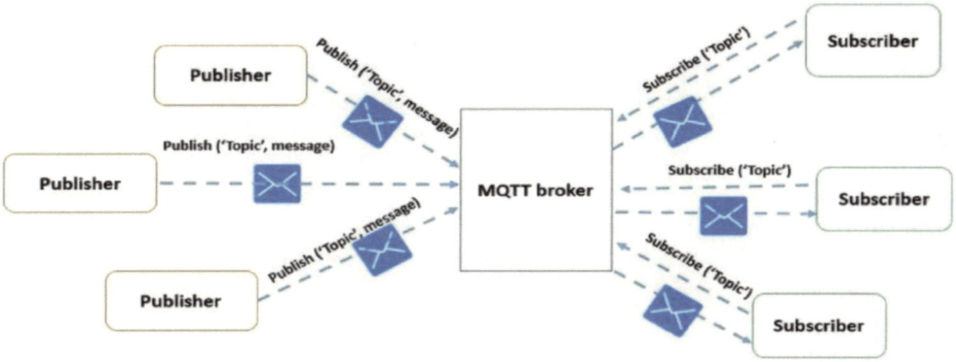

Figure 3.
Broker Publisher/subscriber messaging template.

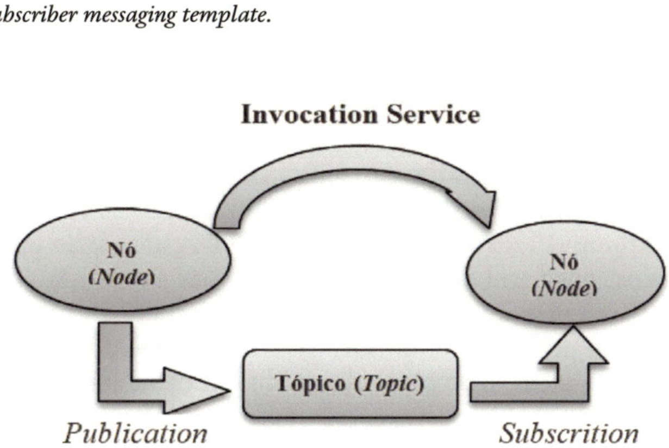

Figure 4.
Publisher/subscriber messaging template.

bit	7	6	5	4	3	2	1	0
Byte 1	Message type				Flag DUP	Level QoS		RETAIN
Byte 2	Remaining width							

Figure 5.
Fixed header of an MQTT message.

Value QoS	Bit	Bit	Description		
0	0	0	Up to once	Shoot and forget	≤ 1
1	0	1	At least once	Delivery with ACK	≥ 1
2	1	0		Guaranteed delivery	1
3	1	1	Reserved		

Figure 6.
QoS levels.

messages exchanged between broker and clients, using predefined topic identifiers and short names of topics in addition to a short message format [7].

According to [8], the PDU of the MQTT protocol is encapsulated by the TCP protocol, that is, the MQTT header and data are sent in the TCP data area (**Figure 5**). In this way the MQTT protocol messages have a fixed header (**Figure 5**) composed of two bytes, where the first byte contains the field that identifies the type of the message, as well as the markers (DUP, QoS level and RETAIN). There is a version of MQTT, called MQTT-SN (MQTT Sensor Network), where your PDU is encapsulated by the UDP protocol, which, in turn, is encapsulated by the IP or the 6LowPAN protocol. One of the main differences between two standards, in addition to the network layer they focus on, is the simplification of messages exchanged between broker and clients, using predefined topic identifiers and short names of topics in addition to a short message format [5].

As can be seen in **Figure 5**, "byte 1" is responsible for four fields:

- DUP (Duplicate delivery): Marker that occupies bit 4 and is activated when the server tries to resend messages of type PUBLISH, PUBREL, SUBSCRIBE or UNSUBSCRIBE (**Figure 5**).

- QoS: This marker represents the reliability of message delivery, indicating the level of guarantee of delivery of a PUBLISH message. You can have up to three levels of guarantee (**Figure 6**). Level 0 is used by those who publish/send the message at most once and does not check whether the message has reached its destination. This lower level is called "fire-and-forget" and the message can be lost depending on network conditions. In Level 1, called the recognized delivery, who publishes/sends a message at least once and check the delivery status using a status check message. However, if this verification message loses, the server broker can possibly send the same message twice, since there was no confirmation that the message has been delivered. Finally we have the QoS2, called guaranteed delivery due to its complicated process, there may be delays end-to-end larger, but no lost messages at this level. The higher the level of QoS, the greater is the packet exchange. If the loss of messages a problem, a lower level of QoS can be used, resulting in less consumption of available bandwidth and less end-to-end delay, which represents limited networks wired or wireless. To further reduce the use of the band, the UDP can be used instead of TCP, but with reduced guaranteed message delivery [9].

3.2 CoAP: constrained application protocol

The Constrained Application Protocol (CoAP), created by the Internet Engineering Task Force (IETF) working group called restricted RESTful environments (CoRE), has been adapted from HTTP, being optimized for devices with limited processing power and capacity, generally applied to intelligent objects in IoT environments [9]. CoAP acts on the UDP transport layer, specifying a minimum set of restrictions such as POST, GET, PUT and DELETE, with some support for resource storage and discovery of embedded resources.

According to [10], CoAP is a transfer protocol aimed at nodes and restricted networks, being designed for M2M applications, such as home automation. CoAP has four types of messages: Confirmable, Non-confirmable, Acknowledgmente and Reset.

- Confirmable (CON): Are those messages that need confirmation at the destination;

- Non-confirmable (NON): They are those messages that do not require acknowledgment of receipt, being very useful for applications that receive constant readings in a short period of time, where the loss of one or the other message does not affect the process;

- Acknowledgment: These are the messages that confirm receipt of a messages, Confirmable;

- Reset: Its function is to indicate that an NOC or NON message was received, but for lack of some context the same could not be properly processed. It can occur in case a device has restarted and the message sent was not properly interrupted.

The COAP uses the request/response model, where devices act as a client or as a server, supporting service discovery and include Web services such as the Uniform Resources Identifiers (URIs) [9].

The following are the main features of the COAP:

- The Coap message exchanges are transported over UDP, and encoding the same are made in binary format with a 4 byte header (**Figure** 7), followed by a variable width token (0 to 8 bytes);

- It has a binary header UDP-based transport, causing thus less delay and reduced battery consumption during transmission;

- Asynchronous message exchange, allowing smart objects to send information only when the application changes;

- HTTP mapping that allows proxies to provide access to CoAP resources.

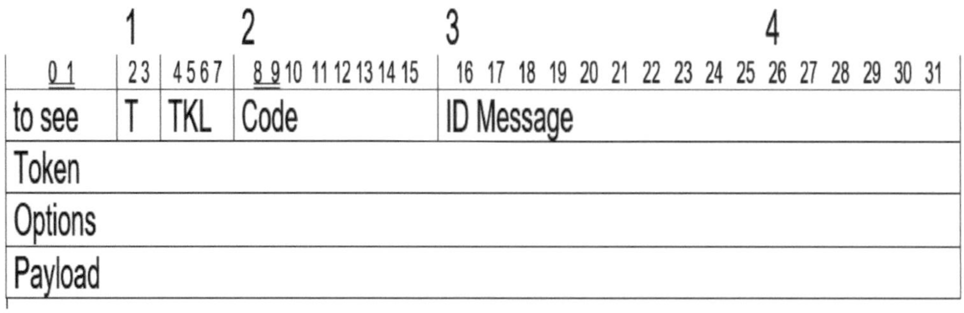

Figure 7.
CoAP message format.

4. Comparison of the protocols

The following will demonstrate a comparison between the protocols presented (MQTT and CoAP), as implementation, data transport, communication standards, reliability and QoS and data security will be analyzed.

4.1 Implementation

Regarding the implementation, the MQTT protocol has a simpler specification in relation to CoAP, thus facilitating customer development. As already mentioned in the 3.2 section, CoAP clients act as HTTP clients but in binary mode, which is simpler than HTTP, but even more complex than MQTT [9].

4.2 Data transport

The MQTT employs a connection oriented communication given by TCP and the CoAP uses UDP. The TCP protocol uses more data to exchange messages between the client and the server in relation to UDP, thus having a higher. Both the MQTT, like CoAP are designed for limited networks like 6LoWPAN (IPv6 over low-power personal area networks). According to [9], if TCP or UDP are not needed, an alternative is to choose the MQTT-SN over 6LoWPAN (IPv6 over low power personal area wireless networks) 4 or even ZigBee, avoiding the complexity of the complete TCP/IP stack. The CoAP It is also designed for limited networks such as 6LoWPAN, in order to maintain short message overload, thus limiting the need for fragmentation that causes significant reduction in the probability of packet delivery.

Regarding the message format, both MQTT and CoAP are suitable to be used in limited bands. Both have a binary message format, different from protocols like AMQP (Advanced Message Queuing Protocol), which uses uses XML formatted messages, in this case requiring the use of more interpreters complex, increasing the hardware requirement.

4.3 Security

One of the main problems to be solved when M2M protocols, is the issue of security [11]. The CoAP protocol is based on DTLS (Datagram Transport Layer Security), so it transfers security handling to the transport layer. Four security modes are allowed:

- NoSec: no DTLS security mechanism is applied;

- PreSharedKey: used with devices that are already pre-programmed with the necessary key switches, where each key has a list of nodes that can communicate;
- RawPublicKey: the device has a pair of asymmetric keys without using a certificate, which is validated by an out-of-band mechanism;
- Certificate: the protocol makes use of the DTLS with an X.509 certificate, the device also has a list of known roots.

According to [12], MQTT security as well as CoAP security (**Table 1**) is performed by Transport Layer Security (TLS). In [13] a safe application model for MQTT is proposed, namely SMQTT. This model is based on a lightweight attribute that provides encryption by broadcast, on elliptical curcas. According to the authors, SMQTT was resistant to attacks from known plaintext, known ciphertex and man-in-the-middle.

Table 2 provides a summary of the security modes of the MQTT and CoAP protocols. In the AAA and Integrity field, it refers to Authorization and

Atack Type	Description
Protocol Parsing and Processing of URIs	It is possible to exploit vulnerabilities in the parsing process (process that analyzes an input sequence), to, for example, generate a denial of service attack by inserting text which will result in parser very extensive.
Proxyinge Caching	The proxy is, in itself, a man-in-the-middle, breaking all the security of IPsec and DTLS. Threats are amplified when proxies allow to cache data.
Amplification Risk	Responses in CoAP are generally larger than requests, which can facilitate amplification attacks
IP Spoofing Attacks	Since there is no handshake for UDP, the final node that has network access can perform spoofing to send messages from ACK instead of CON, preventing from retransmission; spoo pretend the entire payload; spoofing of multicast requests; etc
Cross-Protocol Atacks	They involve using CoAP to send attacks to other protocols, to pass through the firewall, for example
Restriction with Nodes	Whether energetic, memory or processing, make it difficult that devices have good entropy. Therefore, it is assumed, that the processes that need entropy, such as calculating keys, do it externally

Source: [11].

Table 1.
Threats to the CoAP – RFC 7252 protocol.

Protocol	Security Modes	AAA e Integrity	Confidentiality
COAP	No Sec PreSharedKey Raw Public Key Certificate	List of Trusted Roots Uses DTLS	AES-CCM
MQTT	Uses DTLS	Field for name and password uses DTLS	Uses DTLS

Source: [11].

Table 2.
Summary of the security modes of the MQTT and CoAP protocols.

Accountability (Authentication, Authorization and Accountability). In the confidentiality field, the encryptions used by the protocol are described.

4.4 Communication standards

The IoT supports some communication standards that can be defined as:

- In the Telemetry standard, information is sent from devices to the cloud, informing possible changes in states;
- In the query pattern, devices send requests to the cloud to collect information;
- In the Commands pattern, Systems send commands to devices so that they can perform specific activities;
- In the Notification standard, Systems send information to devices in order to inform possible changes in the state of the physical world;

As can be seen in **Figure 8**, the pattern Telemetry becomes suitable for the MQTT protocol, because it has a public/subscribed model, which is equivalent to the telemetry standard. CoAP is not suitable for the Telemetry standard because the connection needs from the system (client) to the server, which faces addressing problems such as mobile roaming or NAT [9]. The CoAP protocol has a better performance for the query communication pattern in relation to the MQTT protocol, since it is based on the request/response model (**Figure 9**). The MQTT has a

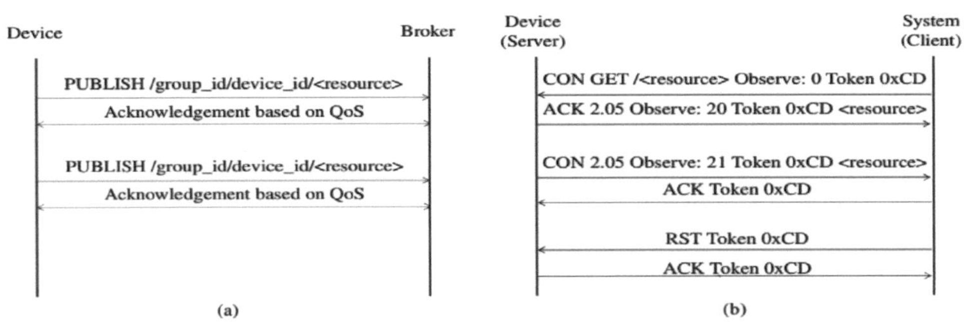

Figure 8.
Telemetry communication pattern example for (a) MQTT, (b) CoAP.

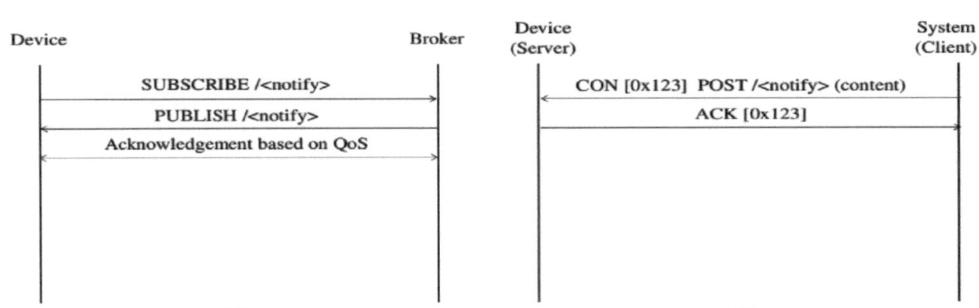

Figure 9.
Example of communication pattern notification for (a) MQTT, (b) CoAP.

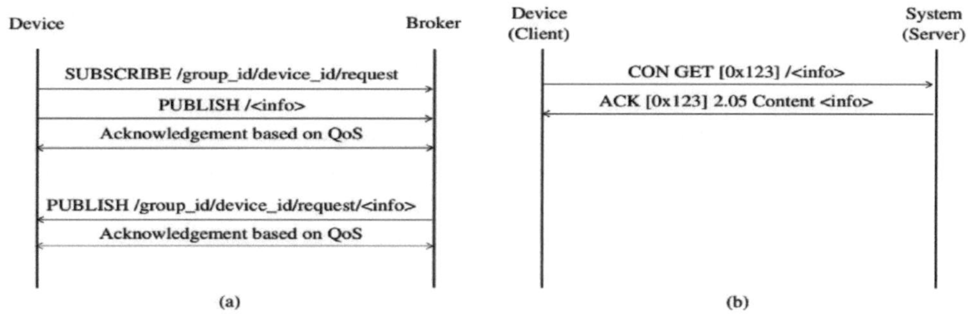

Figure 10.
Communication pattern example query for (a) MQTT, (b) CoAP.

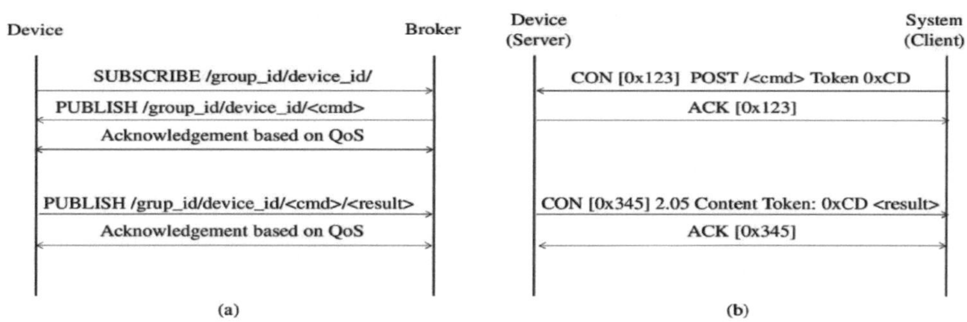

Figure 11.
Communication pattern example command for (a) MQTT, (b) CoAP.

certain difficulty of implementation in the *Consultation* pattern because it has the need to define a response topic for the communication since there is no way for it to be readily constructed (**Figure 10**).

For the *Command* pattern, both protocols face difficulties. CoAP faces the same addressing problems detailed in textit Telemetry and MQTT does not support native result paths, thus requiring a results topic (**Figure 11**).

Finally, in the *Notification* pattern, the CoAP addressing problems, also listed in the Command and Telemetry patterns, are present. On the other hand, the model MQTT publishes/subscribes to the notification architecture, presenting problems only if better flow control is needed for a large amount of data at high rates [4].

5. Conclusion and related works

Based on the information listed in the previous sections, it can be concluded that both protocols (MQTT and CoAP) are considered for use in restricted environments and on devices with battery, processor and limited memory. However, although the two protocols were designed for application in limited environments, the MQTT exchange protocol has the following advantages over CoAP:

- The transport with small overhead makes MQTT an interesting solution for networks with resource constraints, low bandwidth and high latency;

- The MQTT is more geared for communication "many to many" (using the TCP/IP protocol (**Table 3**)), since the COAP is more geared for communication "one to one" for information transfer between client and server, using the UDP/IP protocol;

- Because the MQTT protocol has an exchange of messages based on the publish/subscribe model (**Table 3**), decoupling the sender and receiver from the message both in space and time. Thus the sensors that produce the data do not need to know the identity of the clients who are interested in that information. While in CoAP it is the opposite, the protocol requires the identification of both parties;

According to [14], Due to these characteristics, and mainly the characteristic of being a protocol of communication "many to many", the MQTT protocol, has a greater relevance and use in the existing researches that use scenarios where the number of devices communicating is great. The protocol is used in systems that seek to monitor industrial environments, comparative performance analysis with other industrial protocols of the Internet of Things and M2M and in situations of latency estimation in communication.

In [15], the MQTT and CoAP protocols were responsible for connecting sensors and controlling devices on channels with low bandwidth and little robustness. In this case, they were used in conjunction with the Narrowband-IoT standard (NB-IoT), which has the characteristic of allowing mobile phone communication to be used by devices with limited capacities.

In [13], the authors analyze the feasibility of using ciphertext policy attribute-based encryption (CP-ABE), to allow the security of IoT devices, using the ´MQTT protocol and its variants SMQTT and SMQTTSN.

The authors in [16], the authors use the Prognostic Health Management (PHM) system to detect anomalies in industrial systems. It is proposed in this way the integration of the PHM system to the industrial environment of IoT, based on MQTT and Cloud Computing in order to allow the assessment of the state of the equipment in real time, thus improving the performance of the PHM.

In [17] a cloud-based architecture based on machine learning, for condition monitoring, fault detection and process optimization in industrial environments. The implemented system uses the Dempster-Shafer Evidence Theory (DSET) (**Figure 12**).

In **Figure 12**, the main components proposed can be seen. In this work, the OPC-UA/MQTT gateway is used to communicate between OPC servers on the automation platform and the CORESYS CLOUD broker.

	MQTT	**MQTT-SN**	**CoAP**
Network Protocol	TCP/IP	Not specified	UDP
Useful data type	Binary	Binary	Binary
Suitable for microcontrollers	Yes	Yes	Yes
Security	SSL/TLS	Not specified	DTLS
Scalability	Simply	Simply	Complex
Network architecture	Broker-based (publishes/ subscribes)	Broker-based, Client/Server, Client/Server	Client/Server (request/response)
Network architecture	Broker-based (publishes/ subscribes)	Topic-based	REST architecture
QoS Options	Yes	Yes	Yes

Table 3.
Comparison table between MQTT and CoAP protocols.

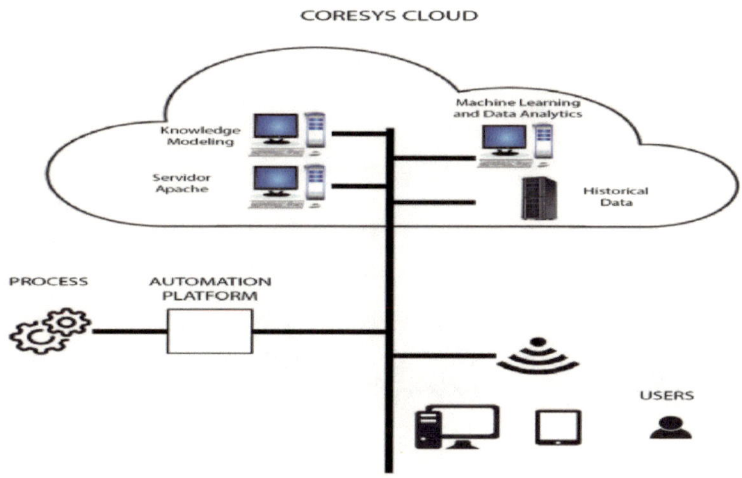

Figure 12.
Framework proposed CORESYS CLOUD. Source: [17].

According to [18], CoreSys-Cloud four services were implemented, namely:

- Machine Learning e Data Analythics (MLDA): Module responsible for learning and analyzing data for state assessment and monitoring;
- Flask: Software used for web application development;
- MySQL database server: responsible for storing the results MLDA evaluation;
- Server MQTT: the MQTT server is responsible for establishing communication between the OPC-UA/MQTT gateway and CoreSys-Cloud.

Cyber security is treated as one of the technological pillars of industry 4.0, which in turn is associated with the protection of software, machines, equipment, network infrastructures and systems. Thinking about it [19] proposed a new approach to protect communication between networked robotic systems, providing authentication and data encryption.The Robot Operating System (ROS) is one of the best robotic software development platforms, offering low-level device control, diverse resources and many useful tools for simulating, visualizing and debugging data, making it very popular with researchers from various robotics fields. Communication in ROS is based on a publish-subscribe system, using the Remote Procedure Call Protocol (RCP) and Extensible Markup Language (XML), with the data sent in clear text over TCP/IP or UDP/IP, without any security mechanism. Based on these aspects, the authors proposed an integration between ROS and the MQTT protocol, using its security features (**Figure 13**).

The authors performed a performance analysis comparing a system without using the security systems offered by MQTT and another system using the MQTT cryptography resources. The results show that the encrypted solution adds negligible delays during communication between clients and servers.

Studies involving the implementation of ROS in industrial environments are gaining more and more evidence. The functioning of ROS is similar to the MQTT protocol in that it works about a publish/subscribe architecture, and use versions ROS of TCP and UDP protocols. Both versions called TCPROS and UDPROS. Due to the use of this architecture, ROS is composed of two elements. The master, which

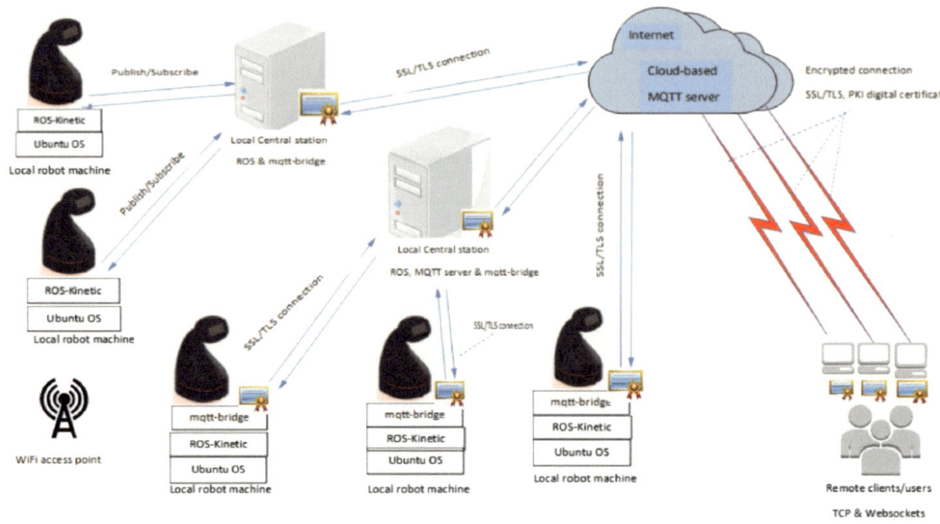

Figure 13.
Proposed architecture. Source: [17].

acts as the MQTT broker, and the nodes, which act as the clients. As o MQTT, ROS controls the data ow through the topics that are coordinated by the master [20]. The main difference between ROS and MQTT, in terms of communication, is the fact that broker has topics with defined typing, that is, a topic will be created with a single type and will receive subscribers and publishers of this type only.

The Authors in [20], presented an implementation of MQTT integrated to ROS, showing the feasibility of using this protocol in the 4.0 industry scenario. [21] cited in [20], evidence the use of ROS in comparison with a traditional solution (**Figure 14**).

In [22] the authors made a comparison between the AMQP (Advanced Message Queuing Protocol) and MQTT protocols, in the context of a smart factory environment, with ROS also being used in some applications that required a more complex and heterogeneous environment.

In [23], The authors proposed os the implementation of internet technology for monitor and control industrial amr robot in industry. Was made a web-based interface for monitor motion and controlling the angle of joint arm robot in a ROS industrial simulation environment, using the mqtt protocol to communicate between the robot and the client, give low latency data transmission.

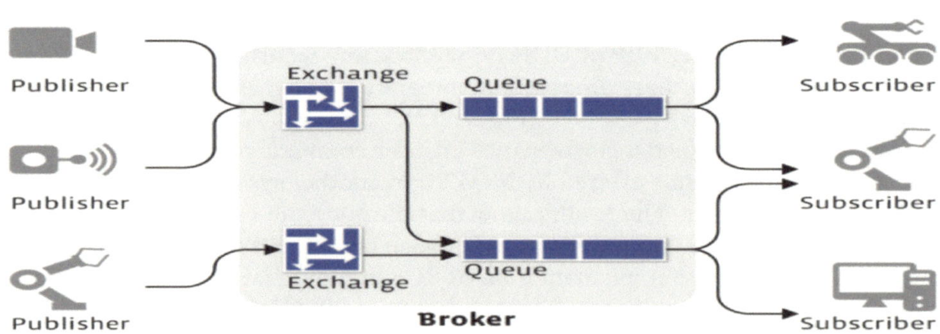

Figure 14.
AMQP architecture.

The MQTT emerges as an excellent alternative for communication between multi robot systems in several other works, as in [16] the the authors conducted a study on the use of MQTT COAP and protocols in Ubiquitous Network Robot Platform (UNR- PF) for the communication of a multi-robot system. In this work, the authors were able to verify that the MQTT protocol is easier to be implemented in the multi robot platform (UNR-PF) than CoAP, in addition to having a higher data transfer rate. Other works related to the use of the MQTT protocol in multi-robot systems are: [14, 24, 25].

Abbreviations

M2M	Machine-to-Machine
IoT	Internet of Things
IP	Internet Protocol
PLC	Programmable Logic Controller
HTTP	Hypertext Transfer Protocol
FTP	File Transfer Protocol
MQTT	Messaging Queue Telemetry Transport
CoAP	Constrained Application Protocol
PDU	Protocol Data Unit
MQTT-SN	MQTT Sensor Network
AMQP	Advanced Message Queuing Protocol
DTLS	Datagram Transport Layer Security
TLS	Transport Layer Security
CP-ABE	Ciphertext Policy Attribute-based Encryption
PHM	Prognostic Health Management
MLDA	Machine Learning e Data Analythics
ROS	Robot Operating System
RCP	Remote Procedure Call Protocol
XML	Extensible Markup Language
UNR-PF	Ubiquitous Network Robot Platform
AMQP	Advanced Message Queuing Protocol

Author details

Edi Moreira M. de Araujo[1*†], Augusto Loureiro da Costa[1†]
and Alejandro R.G. Ramirez[2]

1 Universidade Federal da Bahia, Salvador, Brazil

2 Universidade do Vale do Itajaí, Itajaí, Brazil

*Address all correspondence to: edimmaraujo13@gmail.com

† These authors contributed equally.

IntechOpen

© 2021 The Author(s). Licensee IntechOpen. This chapter is distributed under the terms of the Creative Commons Attribution License (http://creativecommons.org/licenses/by/3.0), which permits unrestricted use, distribution, and reproduction in any medium, provided the original work is properly cited.

References

[1] SCHWAB, Klaus. The fourth industrial revolution. Currency, 2017:1–5.

[2] KAGERMANN, Henning et al. Recommendations for implementing the strategic initiative INDUSTRIE 4.0: Securing the future of German manufacturing industry; final report of the Industrie 4.0 Working Group. Forschungsunion, 2013. https://www.din.de/blob/76902/e8cac883f42bf28536e7e8165993f1fd/recommendations-for-implementing-industry-4-0-data.pdf [Accessed: 01 January 2021]

[3] de Sousa Cardoso, Fábio. UMA ARQUITETURA MULTI-AGENTE PARA COORDENAÇÃO ROBÓTICA. thesis. Rio de Janeiro, 2015. https://w1files.solucaoatrio.net.br/atrio/ufrj-pem_upl/THESIS/1839/pemufrl2015dscfbiodesousacardoso_20170112113019814.pdf [Accessed: 02 January 2020]

[4] MARTINS, Ismael Rodrigues; ZEM, José Luís. Estudo dos protocolos de comunicação MQTT e COaP para aplicações machine-to-machine e Internet das coisas. Revista Tecnológica da Fatec Americana, v. 3, n. 1, p. 24p.-24p., 2015.https://fatecbr.websiteseguro.com/revista/index.php/RTecFatecAM/article/view/41 [Accessed: 3 November 2020]

[5] PTIČEK, Marina et al. Architecture and functionality in M2M standards. In: 2015 38th International Convention on Information and Communication Technology, Electronics and Microelectronics (MIPRO). IEEE, 2015. p. 413-418.. https://www.researchgate.net/publication/308828235_Architecture_and_functionality_in_M2M_standards [Accessed: 6 November 2020]

[6] CAO, Yang; JIANG, Tao; HAN, Zhu. A survey of emerging M2M systems: Context, task, and objective. IEEE Internet of Things Journal, v. 3, n. 6, p. 1246-1258, 2016. https://ieeexplore.ieee.org/abstract/document/7494995/ [Accessed: 10 November 2020]

[7] https://www.emqx.io/blog/mqtt-5-introduction-to-publish-subscribe-model [Accessed: 25 November 2020]

[8] CARISSIMI, Alexandre. Internet das Coisas, middlewares e outras coisas. https://www.researchgate.net/profile/Alexandre_Carissimi/publication/301298394_Internnet_das_Coisas_Middlewares_e_outras_coisas/links/5710f73308aeebe07c023a6e/Internnet-das-Coisas-Middlewares-e-outras-coisas.pdf [Accessed: 05 October 2020]

[9] MAZZER, Daniel; FRIGIERI, E.; PARREIRA, L. Protocolos M2M para Ambientes Limitados no Contexto do IoT: Uma Comparaçao de Abordagens. Inatel. Br, 2015. https://www.inatel.br/smartcampus/imgs/protocolos-para-iot-pt.pdf [Accessed: 10 October 2020]

[10] JIN, Wen-Quan; KIM, Do-Hyeun. Implementation and Experiment of CoAP Protocol Based on IoT for Verification of Interoperability. The journal of the institute of internet, broadcasting and communication, v. 14, n. 4, p. 7-12, 2014. https://www.koreascience.or.kr/article/JAKO201426636276370.page [Accessed: 15 October 2020]

[11] PANDEY, Abhishek; TRIPATHI, R. C. A survey on wireless sensor networks security. International Journal of Computer Applications, v. 3, n. 2, p. 43-49, 2010. https://arxiv.org/abs/1011.1529 [Accessed: 22 October 2020]

[12] DO ESPÍRITO SANTO, Walter; ORDOÑEZ, Edward; RIBEIRO, Admilson. Uma revisão sistemática sobre a Segurança nos Protocolos de Comunicação para Internet das Coisas. Journal on Advances in Theoretical and Applied Informatics, v. 4, n. 1, p. 1-9, 2018. https://revista.univem.edu.br/jad

i/article/view/2482 [Accessed: 28 October 2020]

[13] SINGH, Meena et al. Secure mqtt for internet of things (iot). In: 2015 fifth international conference on communication systems and network technologies. IEEE, 2015. p. 746-751. https://ieeexplore.ieee.org/abstract/document/7280018 [Accessed: 28 October 2020]

[14] BELLAVISTA, Paolo; ZANNI, Alessandro. Towards better scalability for IoT-cloud interactions via combined exploitation of MQTT and CoAP. In: 2016 IEEE 2nd International Forum on Research and Technologies for Society and Industry Leveraging a better tomorrow (RTSI). IEEE, 2016. p. 1-6. https://ieeexplore.ieee.org/abstract/document/7740614 [Accessed: 13 November 2020]

[15] KLYMASH, Mykhailo et al. Designing the Industrial and Environmental Monitoring System based on the Internet of Things Architecture. In: 2018 IEEE 4th International Symposium on Wireless Systems within the International Conferences on Intelligent Data Acquisition and Advanced Computing Systems (IDAACS-SWS). IEEE, 2018. p. 187-190. https://ieeexplore.ieee.org/abstract/document/8525545 [Accessed: 13 November 2020]

[16] AYAD, Soheyb; TERRISSA, Labib Sadek; ZERHOUNI, Noureddine. An IoT approach for a smart maintenance. In: 2018 International Conference on Advanced Systems and Electric Technologies (IC_ASET). IEEE, 2018. p. 210-214. https://ieeexplore.ieee.org/abstract/document/8379861 [Accessed: 23 November 2020]

[17] ARÉVALO, Fernando; DIPRASETYA, Mochammad Rizky; SCHWUNG, Andreas. A cloud-based architecture for condition monitoring based on machine learning. In: 2018 IEEE 16th International Conference on Industrial Informatics (INDIN). IEEE, 2018. p. 163-168. https://ieeexplore.ieee.org/abstract/document/8471970 [Accessed: 25 November 2020]

[18] SOUZA, Daniel Silva de et al. Estudo da aplicação de um sistema IoT baseado no protocolo de comunicação MQTT a área da robótica industrial. 2018. http://repositorio.ufu.br/handle/123456789/24847 [Accessed: 20 December 2020]

[19] MUKHANDI, Munkenyi et al. A novel solution for securing robot communications based on the MQTT protocol and ROS. In: 2019 IEEE/SICE International Symposium on System Integration (SII). IEEE, 2019. p. 608-613. https://ieeexplore.ieee.org/abstract/document/8700390 [Accessed: 20 December 2020]

[20] Daniel Zolett; Alejandro R.G. Ramirez. Desenvolvimento de uma Interface de Monitoração Remota para o Sistema Robótico ROBIX, Integrando o Protocolo MQTT e o ROS. Computer on the Beach. September 2020. DOI: 10.14210/cotb.v11n1.p405-412

[21] Carol Martinez, Nicolas Barrero, Wilson Hernandez, Cesar Montaño, and Iván Mondragón. Setup of the yaskawa sda10f robot for industrial applications, using ros-industrial. In Advances in Automation and Robotics Research in Latin America, pages 186–203. Springer, 2017.

[22] Bezerra, D., Aschoff, R. R., Szabo, G., e Sadok, D. (2018, November). An IoT protocol evaluation in a smart factory environment. In 2018 Latin American Robotic Symposium, 2018 Brazilian Symposium on Robotics (SBR) and 2018 Workshop on Robotics in Education (WRE) (pp. 118-123). IEEE.

[23] Atmoko, R. A., e Yang, D. (2018, August). Online monitoring e controlling industrial arm robot using mqtt protocol. In 2018 IEEE

International Conference on Robotics, Biomimetics, and Intelligent Computational Systems (Robionetics) (pp. 12-16). IEEE.

[24] BOTTONE, Michele et al. Implementing virtual pheromones in bdi robots using mqtt and jason (short paper). In: 2016 5th IEEE International Conference on Cloud Networking (Cloudnet). IEEE, 2016. p. 196-199. https://ieeexplore.ieee.org/abstract/document/7776601 [Accessed: 20 December 2020]

[25] BHAVSAR, Pritesh; PATEL, Sarosh H.; SOBH, Tarek M. Hybrid Robot-as-a-Service (RaaS) Platform (Using MQTT and CoAP). In: 2019 International Conference on Internet of Things (iThings) and IEEE Green Computing and Communications (GreenCom) and IEEE Cyber, Physical and Social Computing (CPSCom) and IEEE Smart Data (SmartData). IEEE, 2019. p. 974-979. https://ieeexplore.ieee.org/abstract/document/8875263 [Accessed: 20 December 2020]